D0871734

APPLICATIONS OF LASER RAMAN SPECTROSCOPY

APPLICATIONS OF LASER RAMAN SPECTROSCOPY

STANLEY K. FREEMAN

Special Scientific Advisor
to the Chairman of the Board

Group Leader—Instrumentation & Analysis
International Flavors & Fragrances
Union Beach, New Jersey

A WILEY-INTERSCIENCE PUBLICATION

JOHN WILEY & SONS, New York · London · Sydney · Toronto

Library of Congress Cataloging in Publication Data:

Freeman, Stanley K.
Applications of laser Raman spectroscopy.

"A Wiley-Interscience publication."
1. Raman spectroscopy. 2. Lasers. I. Title.

QC454.R36F7 547'.34'64 73-12688
ISBN 0-471-27788-6

Printed in the United States of America

10 9 8 7 6 5 4 3 2 1

To

B. M. F.

and

M. E. F.

PREFACE

Observation of the first Raman spectrum more than forty years ago heralded the introduction of a sorely needed technique for structural studies of organic molecules. It was exceedingly difficult to generate an infrared (IR) spectrum. Mass spectroscopy (MS) was used as an analytical tool only, and techniques such as nuclear magnetic resonance (NMR) and electron paramagnetic resonance (EPR) were decades away. During the 1930's and 1940's numerous papers were published on Raman spectroscopy but this discipline was eclipsed in the mid-forties by the advent of moderately priced commercial IR instruments capable of readily recording good quality spectra. Until quite recently Raman spectroscopy was relegated to a very minor role in the scientist's arsenal. However, now it is enjoying a renaissance because the laser has replaced mercury arc sources commonly employed prior to 1968. Essentially all the work presented in this book refers to laser Raman spectroscopy. In order to reestablish itself in the face of a battery of other useful spectral methods the uniqueness as well as the complementary aspects of Raman spectroscopy must be made evident. In my opinion the ancillary value of Raman spectroscopy vis a vis IR spectroscopy has been overly emphasized and consequently most investigators rarely think of using the Raman effect as a general, independent technique. There is a need to demonstrate its value in chemistry and allied fields and it is hoped that this book, based on many and diverse applications of Raman spectroscopy, will fill this need.

Because of its growing acceptance, I have used Wiswesser Line Notation (WLN) on an equal footing with chemical names listed in the tables. Most of the names are based upon Chemical Abstracts nomenclature which, unfortunately, is subject to continuous change. William J. Wiswesser has assigned WLN to approximately 500 compounds; an alphanumeric listing appears in the index.

The reader's attention is directed to a relatively recent change in Chemical Abstracts terminology that has been followed in this book. With reference to geometrical isomerism encountered in molecules containing ethylenic double bonds, the terms "Z" and "E" replace "*cis*" and "*trans*", respectively.

Union Beach, New Jersey
October 1973

Stanley K. Freeman

CONTENTS

APPLICATIONS OF LASER RAMAN SPECTROSCOPY

Chapter 1

GENERAL CONSIDERATIONS

1.1 COMPARISON OF SPECTRAL DISCIPLINES

When one compares Raman, IR, UV, NMR, and mass spectra of a particular compound, the information derived from each method can be unique, complementary, or redundant. Obviously, the degree of data overlap will depend on the nature of the compound under examination. Some salient features of these spectral techniques appear in Table 1.1. One of the purposes of this book is to show in what respect Raman spectroscopy can serve as an independent or a more convenient source of information vis à vis the other spectral disciplines.

1.2 THE RAMAN EFFECT

Light is regarded as both wavelike and corpuscular at the same time. It may be described in terms of its wavelike aspect according to Figure 1.1. The transmission of light through space is encompassed by fluctuating electric and magnetic fields at right angles to each other, their intensities varying sinusoidally with time. Considering the electric vector (Figure 1.2), the distance between any pair of equivalent points is the wavelength, λ, of the radiation which is related to the frequency of the wave motion, ν, by the velocity of a light wave, c. In the corpuscular, or particle, theory light is considered as a stream of energy packets, called photons. Planck's constant, h, relates the energy of the photons, E, of the particle theory with the frequency of the wave theory:

$$E = h\nu$$

TABLE 1.1. Comparison of Spectral Disciplines

Instrumental Technique	Minimum Sample Size (g)	Powders	Single Crystals	Neat Liquids	Aqueous Solutions	Gases[a]	Polymers, Fibers
Raman	10^{-6}	very simple	very simple	very simple	very simple	difficult	simple
Infrared[c]	10^{-6}	simple	very difficult	very simple	difficult	simple	very difficult
Proton NMR	2×10^{-5}	not possible	not possible	simple	very simple	very difficult	very difficult
Mass	10^{-10}	not possible	not possible	very simple	difficult	very simple	not possible
Ultra-violet	10^{-6}	not possible	not possible	very simple	very simple	simple	not possible

TABLE 1.1. (Continued)

Finger-print	Quanitative Analysis	Subtle Structural Features	"On-The-Fly" GLC	General Analysis	Commercial Availability Ref. Spectra	Approx. Instrum. Cost ($)
excellent	good	good	not possible	excellent	poor[b]	15,000 25,000
excellent	good	good	fair	excellent	excellent	7,000- 15,000
fair	good	excellent	not possible	excellent	fair	20,000
good	poor	poor	excellent	good	good	40,000
poor	excellent	poor	poor	poor	excellent	5,000- 12,000

[a] Mol. wt. ∿ 100

[b] Situation improving

[c] Dispersion instruments

[d] Low resolution instruments

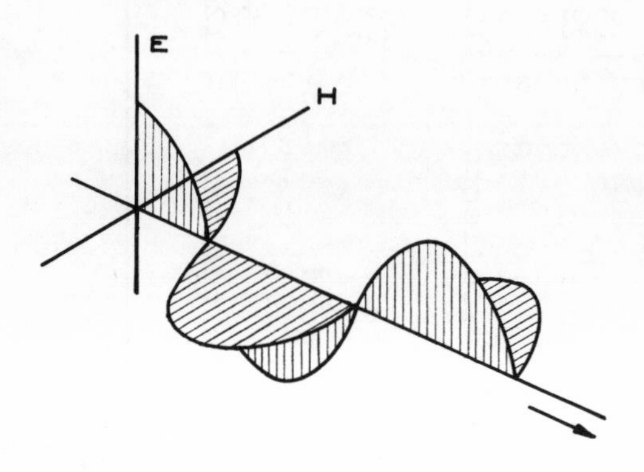

Figure 1.1. Electric (E) and magnetic (H) fields for a plane light wave.

Another quantity used to characterize a light wave is the reciprocal wavelength, known as the wavenumber, $\bar{\nu}$ (cm^{-1}). Since wavenumber is proportional to frequency, the plain symbol ν commonly is employed to denote wavenumber as well as frequency. It will be obvious from the text whether the symbol ν refers to frequency or to wavenumber. The Raman range covers the same region as the so-called middle-to-far IR, falling between visible and microwave light (Table 1.2). IR spectra are plotted as absorption (logarithmic) versus cm^{-1}; Raman as emission (linear) versus cm^{-1} (Figure 1.3). Note that IR bands point down, while Raman bands point up.

Light incident upon a molecule can interact with the molecule by either absorption or scattering. About 50 years ago Smekal (1) predicted the existence of a light scattering phenomenon which was demonstrated by Raman (2) a few years later in 1928. It differed from the previously known radiation scattering effects for particles (Tyndall) and for molecules (Rayleigh). In the Raman effect, photons of the incident radiation interact with molecules of the sample. If the particulate model of light is used, that is, considering a light beam to be composed of a stream of photons of energy $h\nu$, light scattering may be viewed as collisions of these photons with a molecule. Raman spectra are observed when visible light is scattered inelastically by molecules in solids, gases, or liquids. Rayleigh scattering is an elastic process--the photons neither gain nor lose energy in their collisions with molecules. Raman effects are relatively inefficient processes; about 10^{-3} of the intensity of the incident exciting radiation appears as Rayleigh scattering and approximately 10^{-6} as Raman scattering.

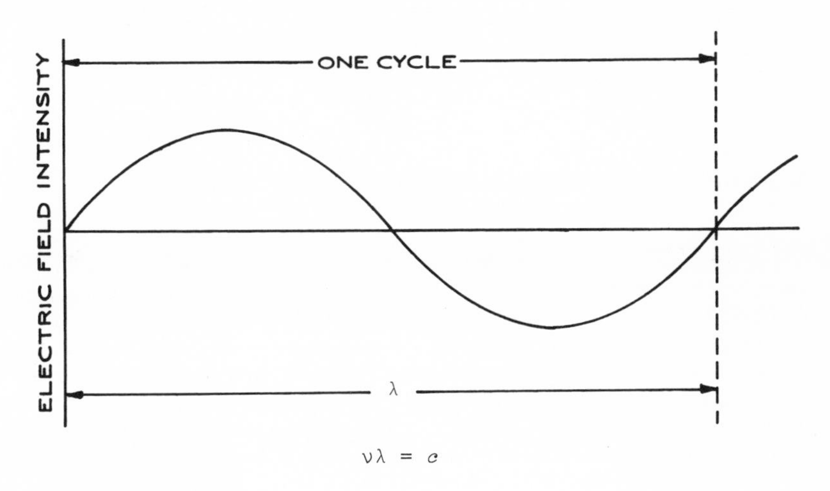

λ = Wavelength

ν = Frequency of oscillation; the number of cycles per second (Hertz)

c = Velocity of light (3×10^{10} cm/sec)

Figure 1.2. The electric vector of a plane light wave.

Figure 1.4 schematically portrays the energy diagram for the Raman effect. The interaction of a photon with a molecule in the ground state ($V = 0$) may momentarily raise the molecule to a higher energy level, or virtual state, that is not stable at room temperature. If the molecule leaves this unstable level it can scatter, or interact with, a photon and return to the ground state. In this instance the scattered photon has the same energy content as the exciting photon, and Rayleigh scattering occurs. On the other hand, if the molecule falls to an excited vibrational state, such as $V = 1$, the scattered photon's energy is now equal to the energy of the exciting photon minus the difference in energy between the $V = 1$ and $V = 0$ levels (ΔV). The frequency of the scattered photon thus is shorter than that of the incident light, giving rise to what is called a Stokes line. An anti-Stokes line is observed when a molecule in an excited level ($V = 1$) rises to a higher, unstable level by interacting with an incident photon and then returns to the ground state upon scattering a photon.

TABLE 1.2. Spectral Ranges of UV, Visible, IR, Raman, Microwave, and NMR

Region		Approximate Range	
		λ	ν (cm^{-1})
UV	Electronic vibration		
Far		10-200 nm	50,000-1,000,000
Near		200-400	25,000-50,000
Visible	Electronic vibration	400-800 nm	13,000-25,000
Infrared	Molecular vibration		
Near		0.8-2.5 μm	4,000-13,000
Middle		2.5-50	200-4,000
Far		50-1000	10-200
Raman	Molecular vibration	2.5-1000 μm	10-4,000
Microwave	Molecular rotation	0.01-10 cm	0.1-100
NMR	Nuclear precession	10-100 cm	0.01-0.1

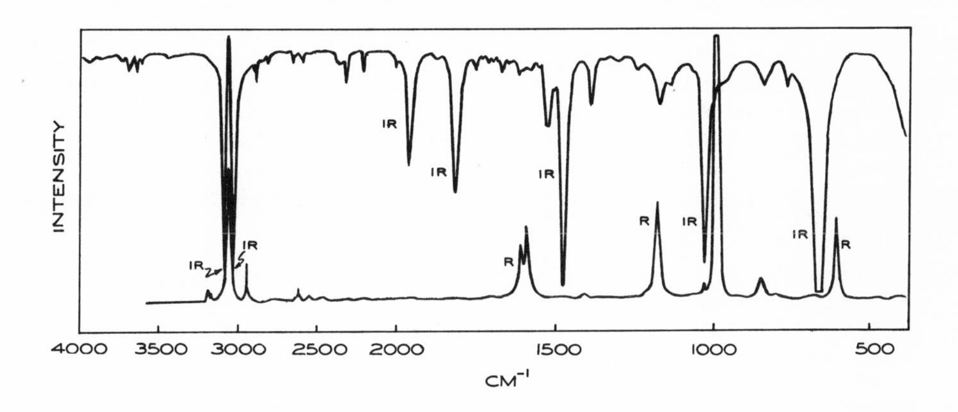

Figure 1.3. Raman and IR spectra of benzene. Note that Raman (R) bands point up and infrared (IR) bands point down.

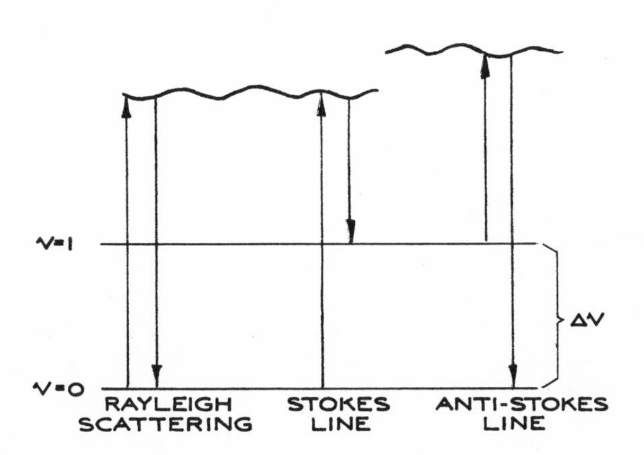

Figure 1.4. Energy level diagram for the Raman effect.

The energy of the scattered photon is equal to the energy of the exciting photon plus the energy difference between the levels $V = 1$ and $V = 0$. In this instance, the observed spectral line has a higher frequency than the incident light. At ambient temperature most molecules are in the ground state; therefore anti-Stokes lines are considerably weaker than Stokes lines. Published Raman spectra usually contain Stokes lines only. In pre-electronic recording days Raman bands were designated as "lines"; this terminology persists today principally in the literature of chemical physicists and physical chemists.

Inasmuch as the energies of the scattered photons are either increased or decreased relative to the exciting radiation, the frequencies of the observed spectral bands are shifted from the frequency of the incident light. These quantized energy increments correspond to the differences in rotational and vibrational levels of the molecule, and in many instances the magnitudes of Raman shifts in frequency correspond exactly to the frequencies of IR absorption. Raman shifts are measured from the exciting line, the latter arbitrarily taken as zero on a relative frequency scale. No matter what the frequency of the laser beam interacting with the sample, the spectrometer will read out the same shift. A ketone displaying an IR absorption band at 1710 cm^{-1} will have a Raman band located at 1710 $\pm$ 3 cm^{-1}, irrespective of the laser excitation frequency employed.

1.3 MOLECULAR POLARIZABILITY

Absorption of infrared radiation occurs only when there is a change in the dipole moment of the molecule during a normal vibration. The intensity of the resulting absorption band is proportional to the square of the change in dipole moment. For a molecular vibration to be Raman active, there must be a change in *induced* dipole moment resulting in a polarizability change of the molecule. To gain a crude idea of the physical meaning of polarizability, consider the behavior of a CCl_4 molecule when placed between oppositely charged plates. The electron cloud of this symmetrical molecule, which does not possess a permanent dipole, becomes distorted and a separation of charge occurs.

Changing the polarity of the plates causes a change in the direction of the cloud; periodic alternations result in a vibrating, induced dipole. Analogously, the fluctuating electric field associated with a light wave impinging on a molecule induces a dipole moment that vibrates with the same frequency as the incident light. Classically speaking, the scattered radiation arises from electromagnetic waves emitted by the vibrating dipole, the intensity being dependent on the polarizability. The polarizability of a molecule is due almost entirely to displacements of the electrons under the influence of the oscillating field of the incident light beam. The induced dipole moment, μ, is proportional to the field strength, E (Equation 1.1). The constant, α, is the polarizability of the molecule.

$$\mu = \alpha E \tag{1.1}$$

Most organic molecules are anisotropic, that is, they are not spherically symmetrical; thus the polarizability may be different in the x, y, and z directions and the expressions in Equation 1.2 can be derived for this general case (3):

$$\begin{aligned} \mu_x &= \alpha_{xx}E_x + \alpha_{xy}E_y + \alpha_{xz}E_z \\ \mu_y &= \alpha_{yx}E_x + \alpha_{yy}E_y + \alpha_{yz}E_z \\ \mu_z &= \alpha_{zx}E_x + \alpha_{zy}E_y + \alpha_{zz}E_z \end{aligned} \tag{1.2}$$

where α_{xx}, α_{xy}, α_{xz},...are proportionality constants, or coefficients, between μ_x and E_x, μ_x and E_y, and μ_x and E_z,...,etc. The polarizability is a tensor because the induced dipole and the field strength are vectors. The polarizability tensor generally is a symmetric one, that is, the off-diagonal components are related as follows:

$$\alpha_{xy} = \alpha_{yx}, \qquad \alpha_{yz} = \alpha_{zy}, \qquad \alpha_{xz} = \alpha_{zx} \tag{1.3}$$

It is a mathematical property of symmetric tensors that there is a special set of axes, x', y', and z', for which only $\alpha_{x'x'}$, $\alpha_{y'y'}$, and $a_{z'z'}$ are different from zero. All terms involving $\alpha_{x'y'}$, $\alpha_{x'z'}$, and $\alpha_{y'z'}$ are equal to zero. Equation 1.2 then reduces to the following form:

$$\mu_{x'} = \alpha_{x'x'} E_{x'}$$
$$\mu_{y'} = \alpha_{y'y'} E_{y'} \qquad (1.4)$$
$$\mu_{z'} = \alpha_{z'z'} E_{z'}$$

where x', y', and z' are three mutually perpendicular directions in the molecule for which the induced dipole moments are parallel to the perturbing electric field, in contrast to the arbitary orientation of the x, y, and z set of axes. The x', y', z' axes are the principal axes of polarizability, and the equilibrium molecular polarizability can be visualized by drawing arrows in any direction from a common origin which have lengths proportional to $1/\sqrt{\alpha}$. The heads of the arrows will define a polarizability ellipsoid whose axes are x', y', and z' (3). For a completely anisotropic molecule, $\alpha_{x'x'} \neq \alpha_{y'y'} \neq \alpha_{z'z'}$; an isotropic molecule has the same polarizability in all three directions. A Raman spectrum will result if the polarizability ellipsoid changes in size, shape, or orientation due to molecular vibration. Polarizability ellipsoids for methane, carbon tetrachloride, carbon dioxide, and sulfur dioxide are shown in Figure 1.5 (4).

1.4 THE RESONANCE RAMAN EFFECT (RRE) AND PRE-RESONANCE RAMAN EFFECT (PRE-RRE)

Resonance Raman scattering involves an interaction of vibrational and electronic transitions. The RRE occurs when the exciting frequency falls in the interior of the observable vibrational structure of the electronic absorption band responsible for Raman scattering (5). The intensities of some Raman emissions may be increased as much as 10^6 times those observed when the exciting frequency is far removed from such a band. In some instances this effect can be used to obtain the spectrum of a sample present in low concentrations by selecting an exciting line that lies near the absorption edge of the material. The resonance Raman spectrum of a material also can be obtained with the aid of a rotating sample cell (see p. 49) and a tunable laser. The spinning cell drastically reduces sample absorption of the incident radiation and subsequent thermal decomposition. Resonance Raman bands increase in strength as the exciting frequency approaches the electronic absorption maximum, attaining optimum intensity when the laser frequency coincides with the absorption maximum. The pre-RRE may occur when the exciting frequency lies close to, but not actually within, the absorption band due to the vibrations in the electronic excited state (6). In the gas phase, excitation producing the RRE usually produces fluorescence, but in solution, liquid, or solid the fluorescence often is quenched and the RRE is observed.

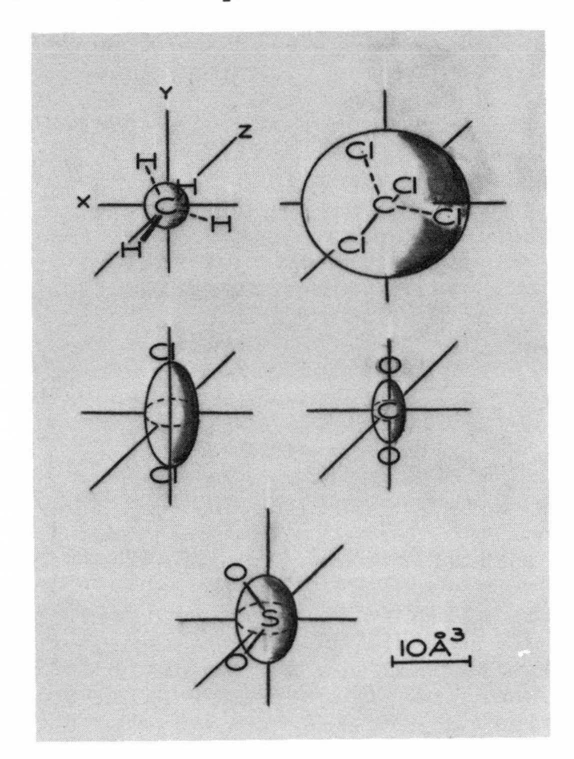

Figure 1.5. Polarizability ellipsoids for some simple molecules drawn to scale (4).

The resonance Raman technique provides a sensitive probe for biological chromophores in their unusual high dilutions (about 10^{-2}-10^{-5} molar). Applications include studies of carotenoid pigments (7), vitamin-A-type molecules (8), rubridoxin (9), hemoglobin (10, 11), oxyhemoglobin (10,12), deoxyhemoglobin (10,12), cytochrome c (13), protoporphyrin IX (heme) (14), and chlorohemin (14).

Classification of the Raman effect, pre-RRE, and RRE can be made in the following way (15):

1. Raman effect: The exciting frequency, ν_o, is far from the frequency of the absorption band ($[\nu_e-\nu_o] \gg 0$) so that the excitation coefficient at the laser line frequency, ε_o, is of the order of 0. The relative intensity of the first overtone to the fundamental is very small ($I_{Ov}/I_f<10^{-2}$). The intensity of Raman lines is described by a ν^4 law (16). (An overtone is a multiple of the fundamental vibration frequency; a combination band is the sum or difference of two or more fundamentals.)

2. Pre-RRE: $[\nu_e-\nu_o] > 0$, $\varepsilon_o \sim 1$ to 100, $I_{Ov}/I_f \sim 10^{-2}$. The ν^4 scattering law requires modification (8).

3. RRE: $[\nu_e-\nu_o] \sim 0$, $\varepsilon_o \sim 10^3$ to 10^4, $I_{Ov}/I_f > 10^{-1}$. The relative intensity of an overtone is approximately proportional to ε_o/n, where n is the vibrational quantum number.

1.5 NONLINEAR RAMAN EFFECTS

A number of phenomena related to the Raman effect occur with very high power laser sources (17). They appear to have little value in chemical analysis and structure determination. Stimulated Raman emission has been applied principally as a new source of radiation at hitherto inaccessible frequencies. Although intense, this effect can be used only over a very short period (~1 nsec). The extremely weak hyper-Raman effect is observed when scattered radiation appears at 2ν, twice the frequency ν of the exciting radiation.

1.6 VIBRATIONAL SPECTRA AND SELECTION RULES

The total energy of a molecule can be represented as a sum of contributions from electronic, vibrational, rotational, and translational parts. Changes in electronic energy result from the absorption or emission of light quanta in the ultraviolet and visible regions of the electromagnetic spectrum. Absorption or emission of quanta in the mid-infrared region of the spectrum and scattering of visible light by vibrating molecules bring about changes of vibrational energy leading to IR and Raman spectra, respectively. Changes in rotational and translational energies of molecules correspond to the far IR and microwave regions.

A nonlinear molecule of N atoms may be viewed as a number of mass points (nuclei) whose average relative positions in space are fixed by an intramolecular force field. Since three coordinates are required to specify the position of each atom, the molecule possesses $3N$ degrees of freedom that are distributed as 3 rotational, 3 translational, and $3N-6$ vibrational. For a linear molecule only two rotational coordinates are needed because it is unnecessary to specify rotation about the molecular axis. Consequently, linear molecules have $3N-5$ vibrational degrees of freedom. Associated with each vibrational degree of freedom is a vibrational frequency. In view of the ground rules for Raman (induced dipole moment) and infrared (noninduced dipole moment), one would expect that the total predicted number of vibrations for a particular molecule would not appear in each type of spectrum. This is indeed the general case.

The polarizability effects in the linear molecule carbon dioxide (18) during its normal vibrations are presented in Figure 1.6. Of the four modes (3 x 3 - 5 = 4), only three are shown. The fourth, ν_{2a}, is a bending vibration in a plane perpendicular to ν_2. The ν_2 and ν_{2a} vibrations, which have nearly the same energy and therefore the same frequency, are termed "doubly degenerate." The symmetric stretching mode, ν_1, clearly is accompanied by a change in the polarizability ellipsoid, but not in the dipole moment itself. This type of vibration, which takes place with conservation of all symmetry properties, is known as a totally symmetric vibration. On the other hand, no polarizability change occurs during the asymmetric stretching or bending vibrations. Although the molecule does not possess a permanent dipole moment, there are changes in the dipole moment for ν_2 and ν_3. Therefore, ν_1 is Raman "allowed" (Raman active) and IR "forbidden" (IR inactive). The reverse holds for ν_2 and ν_3. There are cases of molecules where some of the normal vibrations are forbidden both in the Raman and IR spectra.

A brief discussion of Dermi resonance is in order at this juncture.

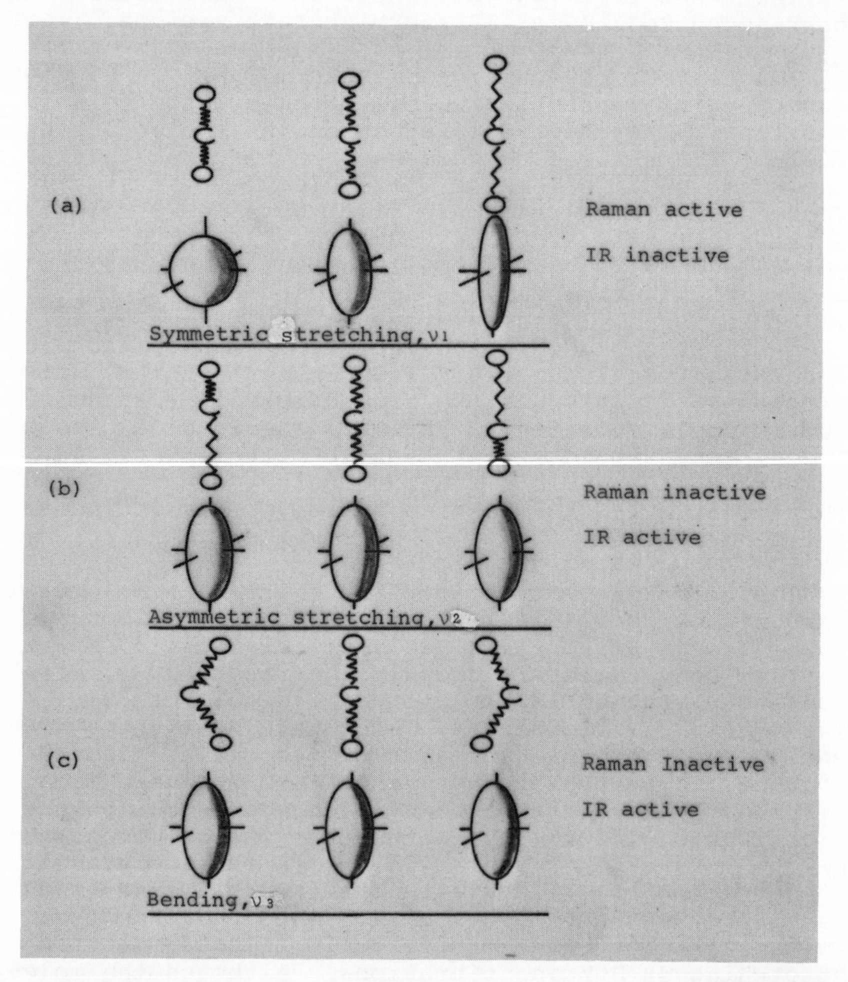

Figure 1.6. Polarizability changes during the vibrations of carbon dioxide (exaggerated). Courtesy of J. Chem. Ed. (18).

Although CO_2 has only one Raman active fundamental, its spectrum contains two strong bands of nearly equal intensity at 1285 and 1388 cm^{-1}. This apparent anomaly may be explained by the fact that the fundamental frequency, ν_1, nearly coincides with the overtone of the bending mode, ν_2, which is expected at about 1334 cm^{-1} (2 x 667 cm^{-1}). Under certain conditions, two vibrational levels belonging to two different vibrations in a polyatomic molecule may have nearly the same energy ("accidental degeneracy"). Such a resonance, or vibrational mixing, occurs between the CO_2 stretching vibrational mode and bending mode overtone. In general, one of two possible effects can occur:

1. If the resonance is close, as in CO_2, two Raman bands of perturbed frequencies occur of about equal intensity, whereas in the absence of vibrational mixing one would expect one strong band (the fundamental) and one weak band (the deformation overtone).

2. If the overtone and fundamental frequencies are not very similar, a significant increase in the weaker band intensity takes place with but little change in spectral position. Ordinarily, an overtone is ten to one hundred times less intense than its fundamental. Overtone bands are encountered more often in IR than in Raman spectra.

It should be noted that Fermi resonance also can occur when a combination band has the same or nearly the same frequency as a fundamental.

The three vibrational modes of sulfur dioxide (18), a simple nonlinear molecule, are illustrated in Figure 1.7. Note that the size, shape, or orientation of the polarizability ellipsoid is altered with each of these vibrations. For the totally symmetric stretching mode, ν_1, the ellipsoid "breathes" with the vibrational frequency and a change in the dipole moment also occurs. Therefore, the vibration is Raman and IR active. The bending vibration of the angular SO_2 molecule does not alter the symmetry of the molecule and so the ellipsoid also swells and contracts with the bending frequency ν_2. The asymmetric stretching mode, ν_3, which is Raman inactive for CO_2, is active for SO_2. Although the polarizability ellipsoid does not change in size or shape, the orientation varies as it rocks to and fro. Both ν_2 and ν_3 modes produce changes in dipole moment. In summary, there are three Raman active and three IR active vibrations for SO_2 with coincidences in the observed frequencies.

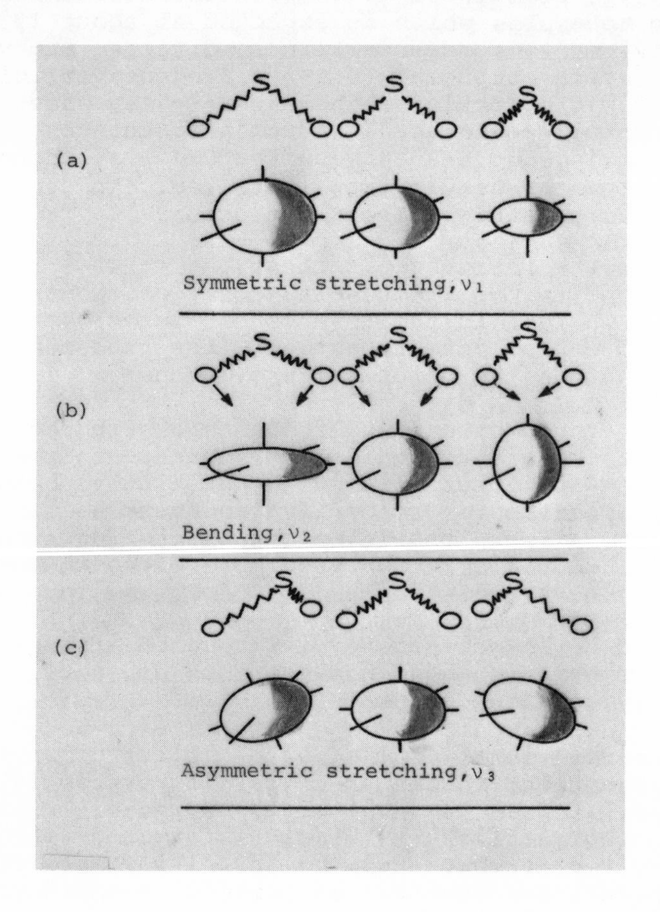

Figure 1.7. Polarizability changes during the vibrations of sulfur dioxide (exaggerated). Courtesy of J. Chem. Ed. (18).

It can be seen, then, that a particular type of vibration will lead to the appearance of a Raman band if the molecular polarizability changes during the vibration. Furthermore, the band intensity depends upon the extent of this polarizability change, while, by contrast, the strength of an IR band is proportional to the magnitude of the dipole moment change. Usually, the IR counterpart of an intense Raman band is manifested as a weak IR band, and vice versa. Since most organic molecules possess little or no symmetry, fundamental vibrations generally will be active both in Raman and IR. It is not within the scope of this book to cover the classification of molecules, which includes symmetry elements, symmetry operations, point groups, and group theory. Library shelves abound with excellent books dealing with this topic (see, e.g., References 19-21) and the reader is advised to spend time on the subject in order to gain a broader view of vibrational spectroscopy. Application of group theory, symmetry elements, and the like however, is limited to relatively simple molecules.

1.7 "MUTUAL EXCLUSION RULE"

Generally, when a molecule has a center of symmetry (inversion center), there is no coincidence between the Raman and IR spectra. A molecule possesses a center of symmetry if a straight line drawn from every atom through the center of the molecule and continued in the same direction encounters an equivalent atom equidistant from the center. Some examples are CO_2, benzene, pyrazine, and (E)-1,2-dichloroethylene (Figure 1.8). In these cases Raman and IR bands will not be observed at the same frequencies if they originate in fundamental vibrations, that is, excluding overtones and combination bands (see p.12). Vibrations asymmetric to the center of symmetry usually will not appear in the Raman spectrum; symmetric vibrations will be IR inactive. This center of symmetry rule is not inviolable, since some breakdown may take place when spectra are recorded on liquids or solids, when molecular complexity increases, or when Fermi resonance (see p.14) occurs. A few examples, showing the utility of the mutual exclusion rule appear in 1.7.1-1.7.5.

1.7.1 Polyethylene Sulfide (PES)

Based on IR measurements and X-ray crystallography, it

has been proposed (22) that PES contains a succession of $(-CH_2-CH_2-S-CH_2-CH_2-S-)$ repeat units. The conformations about the C-C, C-S, S-C, C-C, C-S, and S-C bonds respectively, are *trans*, *gauche* (right), *gauche* (right), *trans*, *gauche* (left), and *gauche* (left). This model for PES has a center of symmetry, but if the material existed in helical form, similar to polyethylene oxide, it would not possess an inversion center. Taking into account the presence of impurities or irregular structures, comparison of the Raman and IR spectra of PES reveals only two coincidences (23). Thus, the Raman spectrum is in accord with the structure derived from IR and X-ray methods.

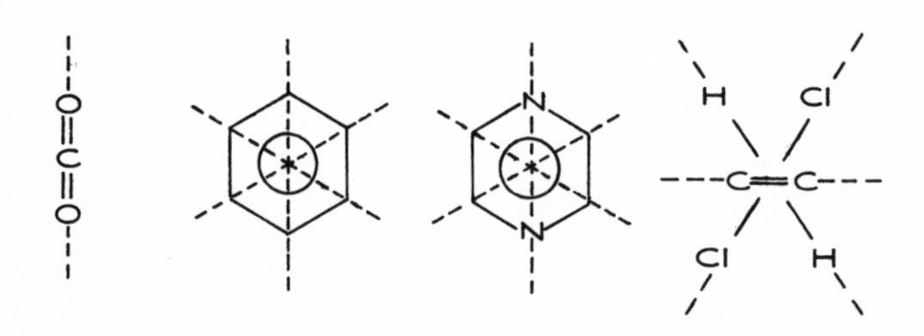

Figure 1.8. Center of symmetry.

1.7.2 Uracil Derivatives

The large number of noncoincidences in the Raman and IR spectra of crystalline 1-*N*-substituted uracils (I) indicates that strong hydrogen bonding prevails (24). The vibrations in the two molecules of the centrosymmetric dimer are coupled, giving symmetric (Raman) and asymmetric (IR) combinations for each coupled monomer vibration.

I

On the other hand, the Raman and IR spectra of 1,3-dimethyl uracil, where hydrogen bonding cannot occur, show the same fundamental bands.

1.7.3 Bicyclopropyl

A good match is obtained between the vibrational frequencies in the Raman and IR spectra of liquid bicyclopropyl, but little spectral coincidence is evident in the crystalline state (25). Thus, the centrosymmetric form (*trans*, II) exists in the solid state, a fact also deduced from electron diffraction studies (26).

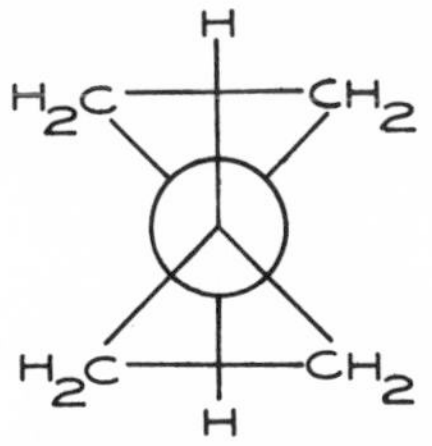

II

Examination of the Raman and IR spectra of 1,1-dimethyl bicyclopropyl (25) indicates that only the rotamer possessing a center of symmetry is present in the liquid state. It may be assumed that intramolecular rotation is restricted by steric hindrance between the two methyl groups. Interestingly, vibrational spectral data show that the isoelectronic 1,1-bisaziridyl exists solely in the *trans* conformation in all states of aggregation. Apparently, repulsion between the nitrogen lone-pair electrons stabilizes the centrosymmetric conformer. This behavior is in contrast to that of hydrazine (27) and its simple derivatives, where the sole existing rotational isomer adopts the *gauche* form having a dihedral angle of approximately 90°.

1.7.4 Diphenylethanes and Diphenylbutanes

Comparison of the solid state Raman and IR spectra of bibenzyl, 1,2-bis-(*p*-chlorophenyl) ethane, α,α'-dimethyl bibenzyl, and 2,3-bis-(*p*-bromophenyl) butane shows only a few coincident bands (28) that probably arise from accidental degeneracy. Therefore, these substances may be presumed to exist in the *trans* conformation in the solid, a conclusion previously supported by X-ray crystallographic evidence (29).

1.7.5 Cyclobutane Photodimers

The mutual exclusion rule has been successfully applied in determining the stereochemistry of some cyclobutane photodimers (Figure 1.9, Table 1.3) (30). The cyclobutane ring of dimers with a head-to-tail (*anti*) or head-to-head configuration exists in planar or puckered conformations, depending on the nature of ring substituents.

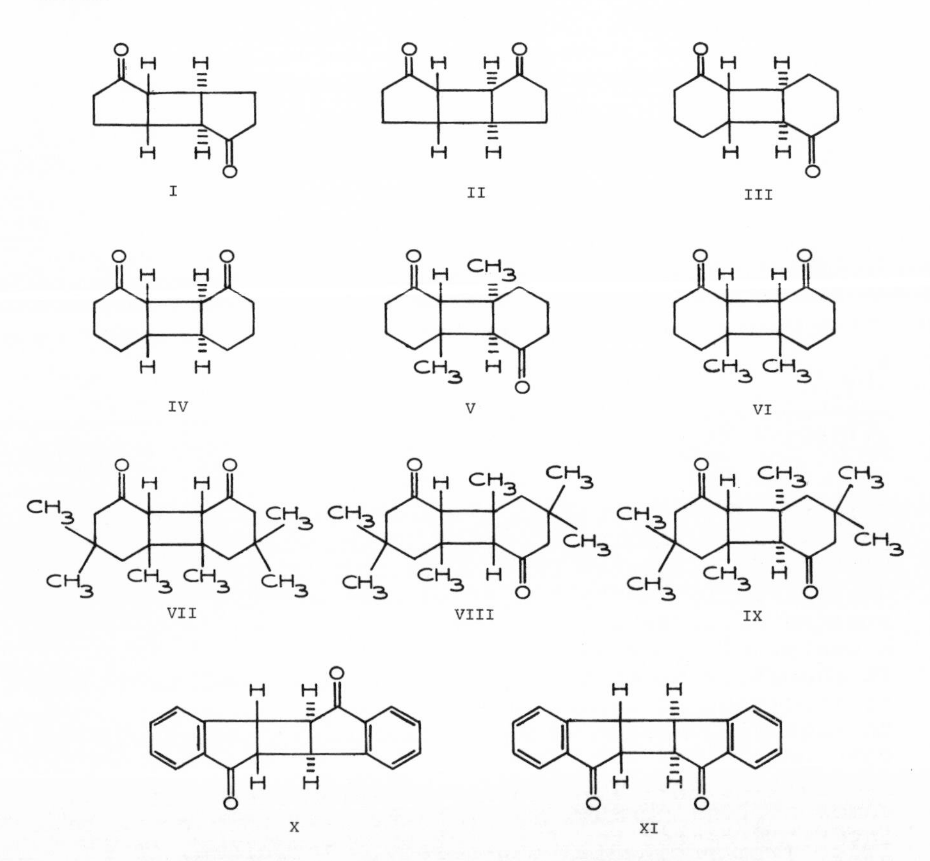

Figure 1.9. Cyclobutane photodimers.

TABLE 1.3. Comparison of the Raman and Infrared Frequency Coincidences[a]

Head-to-Tail Photodimer				Head-to-Head Photodimer			
Compd	*R*	*IR*	*C*[a]	*Compd*	*R*	*IR*	*C*
I	24	24	6	II	31	36	26
III	24	46	10	IV	33	42	24
V	34	34	7	VI	41	46	28
X	26	42	10	XI	24	43	16

Compd	*R*	*IR*	*C*
VII	37	42	23
VIII	41	48	16
IX	33	41	9

[a] R, IR, and C denote Raman lines, infrared peaks, and coincidences, respectively, of compounds listed in Figure 1.9.

Only the *anti* dimer with a planar cyclobutane ring possesses an inversion center. Comparison of the Raman and IR spectra of I shows only six noncoincidences, while twenty-six of the Raman bands in the spectra of II have IR counterparts. Thus, I possesses a center of symmetry, in agreement with X-ray data. It should be noted that in this type of comparison it is important to establish a definite trend rather than to require a strict mutual exclusion for one of the compounds. For the cyclohexanone dimers, III shows ten coincidences while there are twenty-four for IV; somewhat less persuasive, but still indicative of an inversion center for III. This approach also allows differentiation between the corresponding methyl derivatives V and VI. Seven coincidences occur in the spectra of V whereas twenty-eight appear in VI. The three isophorone photodimers pose a more difficult problem. As the molecule increases in size the spectra become more complex due to shifts in vibrational frequencies through coupling effects between a large number of vibrations of the same symmetry and to accidental degeneracies. We can be reasonably certain in stating that VII does not have a center of symmetry, because twenty-three coincidences are observed. Of the two *anti* dimers, VIII and IX, it would appear that the one with only nine coincidences has the center of symmetry and, therefore, is IX. However, in the absence of VII and IX, the other compound, assigned as VIII, is a border line case (sixteen coincidences) for establishing whether or not it possesses an inversion center. The situation becomes less clear in the final pair. Structure X, with a potential center of symmetry, shows ten spectral coincidences; noncentrosymmetric XI has sixteen. One can rationalize the larger number of coincidences in X compared with I on the basis of the greater opportunity for accidental coincidences as the molecular size increases. On the other hand, several conformers may coexist, that is, a small percentage of molecules in a puckered state may be present. We can conclude that compounds I-VII can, and VIII-XI cannot, be unambiguously determined by applying the rule of mutual exclusion. In many cases where dimers can be compared, the compound possessing a center of symmetry can be confidently assigned. Furthermore, if NMR, dipole moment, or other physical methods indicate that dimerization occurs in the head-to-tail fashion, vibrational spectroscopic comparisons can yield valuable information in completing the structural analyses.

1.8 GROUP FREQUENCIES

Raman spectroscopy historically preceded IR in the study of group frequencies as related to molecular structure, and a number of useful group frequencies has been recognized for many years (31,32). Until recently, Raman spectroscopy has been in the purview of theoretical spectroscopists for elucidating structures of small molecules. Improvements in instrument design, sample handling (see Chapter 2), and the introduction of laser sources have brought about a dramatic change. Spectra of large molecules now can be generated, and group frequency correlations enable the investigator to obtain unique as well as complementary information from a Raman spectrum.

Two types of vibrations are associated with covalently bound atoms: (1) a periodic bond extension and contraction (stretching) and (2) a periodic bending, or deformation. They have been described previously (see Figures 1.6 and 1.7) for the CO_2 and SO_2 molecules and are pictured in Table 1.4. More complex types of vibrations also occur, involving the simultaneous stretching and bending of several bonds linked to the same atom. The frequency of a stretching vibration (ν) depends on the bond strength, or force constant ($K \times 10^5$), and inversely on the masses of the two atoms involved (M_C and M_X); see Equation 1.5.

$$\nu = 1303 \sqrt{K\left(\frac{1}{M_C} + \frac{1}{M_X}\right)} \qquad (1.5)$$

It is found that a C≡C group scatters light near 2200 cm^{-1} and a C=C group near 1650 cm^{-1}. Stretching vibrations of C-F, C-Cl, C-Br, and C-I groups appear near 1200, 700, 550, and 500 cm^{-1}, respectively (see Table 1.5). Since stretching force constants are considerably greater than bending force constants, the lower energy deformation vibration bands appear at lower frequencies, for example, CH_2 rock at about 720 cm^{-1}, CH_2 in-plane bend at about 1450 cm^{-1}, and CH_2 stretch at about 2900 cm^{-1}. The concept of characteristic Raman or IR frequencies arose from vibrational analysis and empirical findings that certain groups of atoms, mainly functional groups, gave rise to bands of characteristic frequency. For example, in the absence of disturbing influences, the characteristic band associated with a C=O stretching vibration would lie at a specific frequency, irrespective of the molecule in which this group is present. But the rest of the molecule <u>does</u> affect the bond force constant and the "constancy" of the group frequency holds only when the molecular environments are similar. This type of effect is used to advantage in interpretive vibrational spectroscopy. Essentially, the presence or absence of a band within a relatively narrow frequency range can be used to infer the presence or absence of a particular grouping. The bands above 1500 cm^{-1} in an IR spectrum are assigned to functional groups, and those absorptions observed below 1500 cm^{-1} comprise the fingerprint region. Some functionalities yield diagnostic IR bands in the fingerprint region, for example aromatic rings, ethers, vinyl, and vinylidene moieties. Raman group frequencies occur above about 400 cm^{-1}.

TABLE 1.4. Vibrations Associated with Covalently Bound Atoms

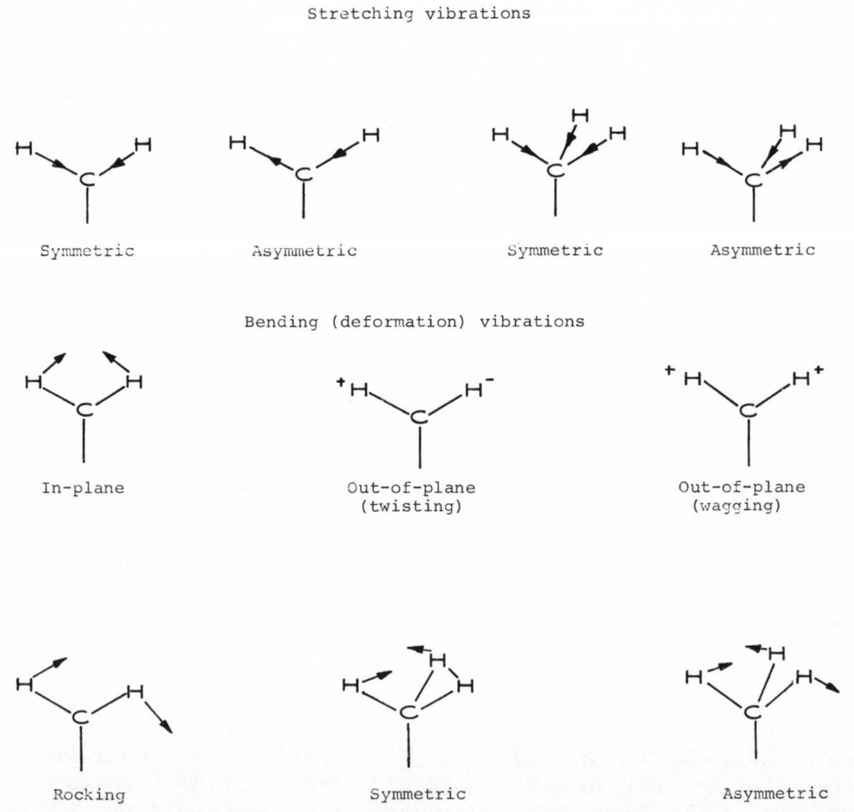

Stretching vibrations

Symmetric | Asymmetric | Symmetric | Asymmetric

Bending (deformation) vibrations

In-plane | Out-of-plane (twisting) | Out-of-plane (wagging)

Rocking | Symmetric | Asymmetric

TABLE 1.5. Stretching Vibrations and CH_2 Vibrations

Stretching Vibrations				CH_2 Vibrations	
Group	cm^{-1} (Approx.)	Group	cm^{-1} (Approx.)	Type	cm^{-1} (Approx.)
C≡C	2200	C-F	1200	Stretch	2900
C=C	1650	C-Cl	700	In-plane bend	1450
C-C	1000	C-Br	550	Rock	720
		C-I	500		

Several excellent books describing IR characteristic group frequencies can be used for those groupings that give moderate to strong Raman bands (e.g. References 33 and 34). Table 1.6 contains a few correlations for the Raman effect.

1.8.1 Vibrational Coupling

The vibration of an X-H bond can be treated as a separable group frequency in those instances where a relatively large mass difference occurs between X and H. However, a characteristic frequency cannot be assigned to groups X-C or C-Y in molecule or group X-C-Y if the masses of X, C, and Y are similar. The X-C and C-Y vibrational modes contain substantial contribution from the stretching of both bonds. When they have nearly the same frequency, nearly complete coupling, or interaction, takes place and the resultant two frequencies are displaced from their original positions by an amount depending on the extent of coupling. Coupling also can take place between two identical groups in the same molecule or even when groups are separated by several atoms. As an example, two carbonyl bands are found near 1815 and 1785 cm^{-1} for diacyl peroxides arising from coupling between the in-phase and out-of-phase vibrations of the C=O groups. The absence of significant coupling between groups in organic molecules makes possible good group frequency data. If this were not the case, interpretation of vibrational spectra would be nearly impossible.

TABLE 1.6. Raman Frequencies (Approximate) and Relative Intensities (Approximate) for Some Functional Groups

Group	ν (cm^{-1}) (Approximate)
NH_2	3300-3500
SH	2550
C≡N	2250
C≡C	2100
C=O	
COCl	1780
COOR	1740
CHO	
unconj.	1725
conj.	1695
CO	
unconj.	1720
conj.	1670
COOH	
unconj.	1670
conj.	1630
C=C (acyclic)	1550-1700
C=C-$\underline{CH}_3$	1375
C-S	600-700
S-S	500
S-S-S	450-500

1.9 DEPOLARIZATION RATIOS

1.9.1 Normal Polarization

Associated with the polarizability are the isotropic, or spherical, part ($\bar{\alpha}$) and the completely anisotropic part ($\bar{\beta}$). Equation 1.6 relates the depolarization ratio (ρ) with $\bar{\alpha}$ and $\bar{\beta}$.

$$\rho = \frac{3\bar{\beta}^2}{45\bar{\alpha}^2 + 4\bar{\beta}^2} \qquad (1.6)$$

While the minimum value of ρ is zero in the special case of a totally symmetric vibration that scatters only polarized radiation ($\bar{\beta} = 0$), a maximum figure of 3/4 is obtained when a vibrational mode completely depolarizes the incident light ($\bar{\alpha} = 0$). Although Raman theory allows predicting whether a vibration will give rise to a highly polarized or a highly depolarized line, ρ values lying between these extremes must be gained experimentally. When ρ is less than about 0.2, the $\bar{\alpha}$ term probably dominates, that is, a more symmetric change in the polarizability ellipsoid occurs, which is consonant with a low degree of distortion during the vibration. On the other hand, the $\bar{\beta}$ term becomes more important when ρ is greater than about 0.2. In this instance, the off-diagonal component of the polarizability ellipsoid, which represents the distortion of the ellipsoid during vibration, assumes more influence.

The depolarization ratio, or the degree of depolarization, associated with a particular vibration is a very important parameter in Raman spectroscopy and should be reported along with the frequency of a band. It has no true counterpart in IR. Prior to the introduction of laser excitation sources, depolarization ratios were used principally to assist in assigning symmetric and asymmetric vibrational modes. Recently, polarization data have been found to serve as subtle probes in sorting out various types of carbonyl (see Chapter 4) and alkene (see Chapter 5) vibrations. An understanding of the basic concept and procedure to determine the degree of depolarization of compounds in the liquid or gaseous state may be gained by referring to Figure 1.10. The polarizability is isotropic, or the same in all directions, for a molecule whose electron density is spherically symmetrical. When the molecule interacts with laser radiation polarized in the XZ plane [Figure 1.10 (a)], the light scattered perpendicular to the incident beam will be polarized nearly exclusively in the YZ plane before entering the spectrometer slit. The emitted radiation is viewed at right angles because an incident quantum produces a dipole moment perpendicular to its direction of motion only. A polarizing analyzer (P) oriented in the YZ plane essentially is transparent only to light in this plane. When the analyzer is rotated through 90°, the YZ plane radiation is blocked and no scattered light passes through the entrance slit of the spectrometer [Figure 1.10 (b)]. No YZ, or depolarized, component is present in the scattered light because a symmetrical molecule does not depolarize the laser beam. The depolarization value for a Raman band is the ratio I_x/I_z, where I_x and I_z are intensities of the scattered radiation polarized in the XY and YZ planes, respectively. Thus, ρ is zero for a molecule whose polarizability is isotropic. The polarizability of an organic compound usually is anisotropic, being different for different directions in the molecule, and some depolarization of the incident polarized light occurs.

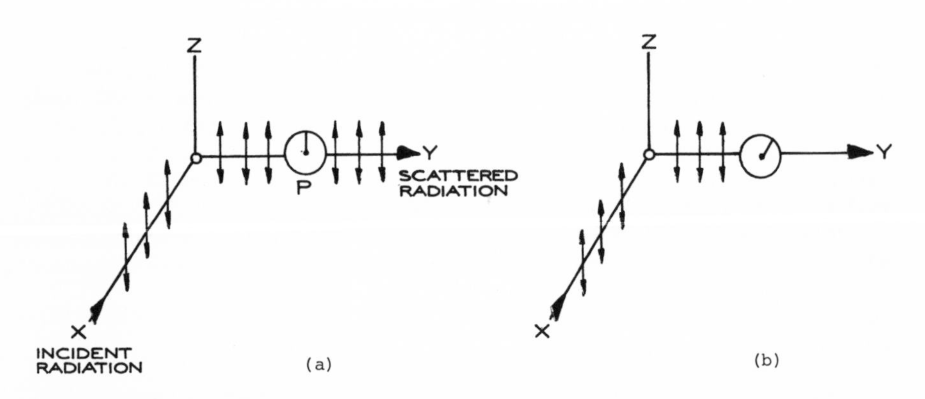

O = Molecule
P = Polarizer

XZ Plane-- Plane of laser radiation
YZ Plane-- Plane of scattered, polarized radiation
XY Plane-- Plane of scattered, depolarized radiation

Figure 1.10. Polarization measurements. (a,b): Totally symmetric molecule reacts with incident polarized light and scatters polarized radiation only. (a): Parallel-oriented polarizer allows passage of scattered light in YZ plane. (b): Perpendicular-oriented polarizer does not transmit YZ-plane radiation.

A parallel-oriented analyzer will permit passage of polarized radiation only, whereas a perpendicular arrangement transmits only the depolarized light [Figure 1.10 (c) and 1.10 (d)].

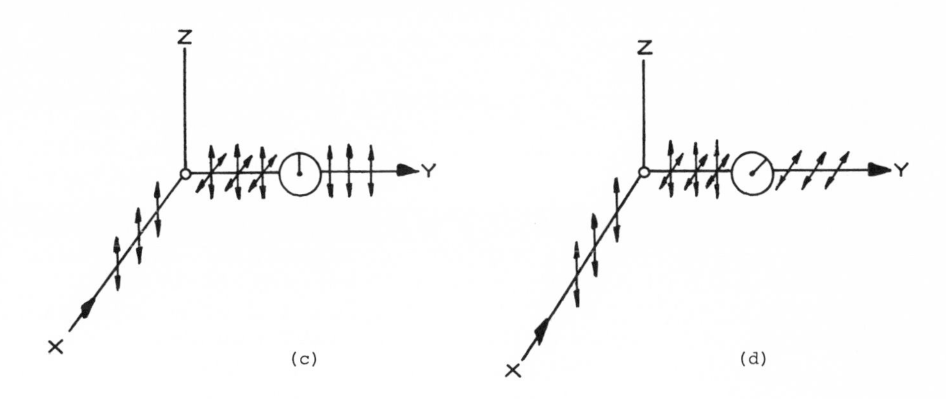

Figure 1.10. Polarization measurements (Continued) (c,d): Asymmetric molecule scatters both polarized and depolarized radiation. (c): Parallel-oriented polarizer transmits polarized radiation (YZ plane) only. (d): Perpendicular-oriented polarizer transmits depolarized radiation (XY plane) only.

The depolarization value, ρ, is obtained by ratioing the band intensities (peak height or peak area) obtained when the polarizer is oriented perpendicular and parallel with respect to the scattered, polarized radiation: $\rho = I_{\perp}/I_{\parallel}$.

Before undertaking a depolarization ratio measurement, the Raman spectrometer should be standardized against the depolarized 218 cm^{-1} ($\rho = 0.75$) and polarized 459 cm^{-1} ($\rho = 0.005 \pm 0.002$) bands of carbon tetrachloride. Theoretically, the depolarization ratio is defined in terms of integrated band intensities, but the ratio of peak heights is adequate. Obtaining the depolarization value of a spectral band is simply and rapidly accomplished in the case of a liquid material. In many modern Raman spectrometers a polarizing analyzer is located between the sample and entrance slits. The region of interest is recorded first with the analyzer oriented parallel and then perpendicular to the electric vector of the exciting light. The ratio of peak heights of the ⊥ and || scans is calculated to gain the ρ value. A commercially available Raman spectrometer (Jeol JRS-Sl) is equipped with a device that automatically performs these operations and records the depolarization ratio on the spectral chart.

Polarization data on gases and oriented single crystals are seldom utilized by the organic chemist. Although values cannot be obtained directly on finely divided crystals or powders, because reflection and refraction at the solid surfaces lead to polarization scrambling, this type of information may be gained on solutions of the materials (35). The depolarization ratio of a Raman band can be determined for polymer films, as in liquids and gases, by measuring $I_{\parallel}$ and $I_{\perp}$.

1.9.2 Inverse Polarization

Equation 1.6 is valid only if the polarizability tensor is symmetric, that is, as defined by Equation 1.3. Based on theoretical considerations, the tensor may become asymmetric when the exciting radiation approaches resonance with an electronic transition of the molecule (5). The polarization can become anomalous ($\rho > 3/4$) for totally symmetric modes under these conditions. The phenomenon of inverse polarization, $\rho = \infty$, occurs only in the case of a nontotally symmetric vibration for which the polarizability tensor is asymmetric, that is, $\alpha_{xy} = -\alpha_{yx}$, and so forth. Such modes are forbidden in nonresonance scattering. Bands exhibiting inverse polarization have been observed in the resonance Raman spectra of hemoglobin and cytochrome c in dilute solutions (36).

1.9.3 Depolarization Ratios and Short-Range Order in Liquid Crystals

Liquid crystals are obtained when most organic substances are melted, forming at that stage of melting when insufficient heat is supplied to cause the transition into the normal isotropic liquid. They are phases that are strongly anisotropic in some of their properties, a characteristic generally found only in solid crystals, although the phases themselves may be as mobile as water. There are three principal types of crystalline liquids: nematic, cholesteric, and smectic. Molecules that form mesomorphous phases are elongated with their long axes parallel.

Raman spectroscopy has been used to investigate the phase transitions in liquid crystals (37-40). Significant spectral changes occur only in the solid nematic phase transition and not in the nematic liquid transition. The results, consistent with the observation of a very small change of latent heat associated with the nematic liquid phase transition (41), suggest that Raman spectra of the nematic and isotropic liquid phases are primarily affected by short-range order. Raman depolarization ratio studies can provide information about the nematic liquid phase transition (40). Molecular motions in the nematic and isotropic phases are different because in the nematic phase only the reorientation around the molecular long axis is possible, while reorientation about the short axis is also permitted in the isotropic phase. Hence, the depolarization ratio rather than the total scattering intensity should reflect the change of short-range order as the phase transition occurs. The results of an investigation of *p*-(*p*-ethoxyphenylazo)-phenyl undecylenate (PPU) have provided information on such a change (40). Polarization measurements were made on the 1145 cm^{-1} CH_2 twisting mode of PPU at 327.6 (solid), 350.2 (nematic), and 377.7°K (liquid). Despite the large decrease in band intensity as the solid to nematic phase transition occurs, the polarized and depolarized spectra are similar in the nematic phase. Since this effect is observed not only in the 1145 cm^{-1} band but in other bands as well, the polarizability of PPU in the nematic phase is very anisotropic. The depolarization ratio reflects the overall reorientation of the molecule and Equation 1.6 is valid only if the molecular orientation is isotropic. Depolarization values of 0.3 and about 1 for PPU in the isotropic liquid and nematic phases, respectively (Figure 1.11), indicate that below the nematic liquid transition temperature reorientation is not permitted around the molecular short axis because of short-range order.

1.10 RAMAN SCATTERING FROM OPTICALLY ACTIVE MOLECULES (CIRCULAR DIFFERENTIAL RAMAN SPECTRA)

A wave of monochromatic plane polarized light may be

considered to be made up of two vectors, corresponding to left and right circularly polarized waves which are similar to left and right handed helices.

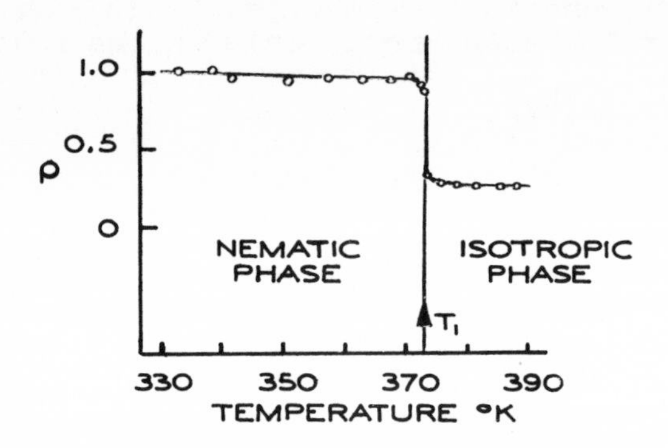

Figure 1.11. The depolarization ratio of the 1145 cm^{-1} band of PPU as a function of temperature (T = nematic liquid transition temperature) (40).

When the waves associated with these helices rotate at different speeds, the plane of polarization is rotated through an angle. Since the speed of a light wave is a function of the refractive index of the medium it traverses, the medium will be optically active if it has different indices of refraction for left and right circularly polarized light (circular birefringence). The change in optical rotation with wavelength gives an optical rotary dispersion (ORD) curve. Not only are the left and right handed helical waves propagated with different velocities in the region of an absorption band, they also are absorbed to different extents. Circular dichroism (CD) is the variation of this absorption difference with wavelength.

Currently available instruments for measuring ORD and CD cover only the visible and near ultraviolet spectral range. As a result, gaining useful data on optical activity is restricted to a relatively small number of molecules, or to regions of large molecules, which absorb light at these frequencies. ORD and CD in the infrared region are several orders of magnitude smaller than in the visible and near-ultraviolet because optical activity depends on the frequency of the exciting radiation. Raman spectroscopy on the other hand, provides a means of obtaining vibrational spectra with visible light, making possible the practical application of the Raman effect for studies of optical activity.

Recently, it has been observed that the intensity of Raman scattering from optically active molecules is slightly different in left and right circularly polarized incident light. The molecular properties responsible for optical activity are the tensors $\underline{G}'$ and $\underline{A}$, giving the electric dipole moments induced by a magnetic field and an electric field gradient, respectively (42-46). Interesting conclusions have been drawn from a consideration of the contribution of scattering through $\underline{G}'$ and $\underline{A}$ to the intensity and depolarization of Raman light (47-51). The scattered intensity, proportional to $\underline{G}'^2$ and $\underline{A}^2$, is only $\sim 10^{-6}$ times that arising from α^2 scattering. A circular intensity differential arises from interference between the polarizability tensor α and the optical activity tensors $\underline{G}'$ and $\underline{A}$. Scattering involving $\alpha\underline{A}$ and $\alpha\underline{G}'$ is about 10^{-3} times the α^2 scattering, but since the latter is the same in right and left circularly polarized incident light, the $\alpha\underline{G}'$ and $\alpha\underline{A}$ scattering can be distinguished from the α^2 scattering by modulating the incident circular polarization at a suitable frequency between right and left, and detecting the modulated components of the scattered light.

The circular intensity differential (CID) is defined by

$$\Delta = \frac{I_R - I_L}{I_R + I_L}$$

where I_R and I_L are the scattered intensities in right- and left-hand circularly polarized incident light, respectively, and Δ is similar to the "dissymmetry factor" in electronic circular dichroism (48). Raman CID was first reported for α-phenethylamine (49) and recently for α-phenylethylisocyanate (48). A relatively large Raman CID couplet was observed in the latter case, consisting of a positive and a negative component associated with two weakly polarized bands at 223 and 313 cm^{-1} (Figure 1.12). The (+) and (-) enantiomers give mirror image couplets.

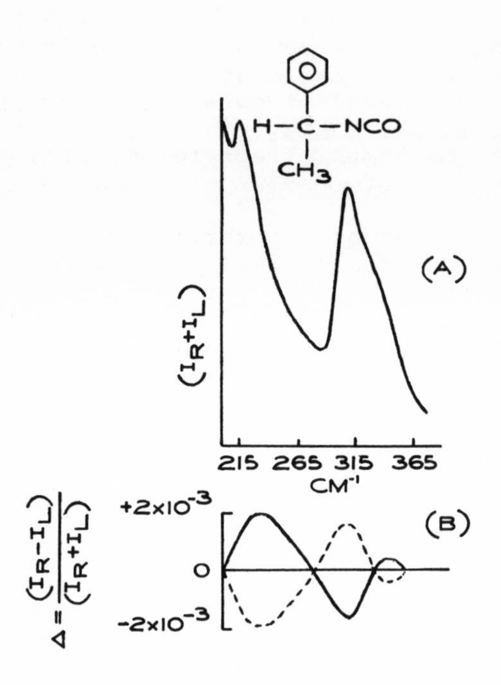

Figure 1.12. Raman (a) and Raman CID (b) spectra of α-phenylethylisocyanate for scattered light linearly polarized parallel to the scattering plane. Solid curve: (-) enantiomer; dotted curve: (+) enantiomer (48).

Raman CIDs have been reported in several monoterpenes, possibly originating in ring vibrations involving asymmetric carbon atoms (51).

These results indicate that a method exists for directly probing the dyssymmetric environment of a functional group exhibiting a characteristic group vibrational frequency. If this phenomenon is confirmed by further experimental data, Raman CID could provide a complementary, and possibly a better criterion than rotary power (which depends on subtle and sometimes unpredictable electronic effects) for correlating absolute configuration.

REFERENCES

1. A. Smekal, Naturwissenschaften 11, 873 (1923).
2. C. V. Raman and K. S. Krishnan, Nature 121, 501 (1928); C. V. Raman, Indian J. Phys. 2, 387 (1928).
3. N. B. Colthup, L. H. Daly, and S. E. Wiberley, *Introduction to Infrared and Raman Spectroscopy* (Academic, New York, 1964).
4. H. A. Stuart, *Molecular Structure* (Springer, Berlin, 1934).
5. G. Placzek, *Rayleigh and Raman Scattering*, UCRL Trans. No. 526 L from *Handbuch der Radiologie*, E. Marz Leipzig (Ed.), Akademische Verlags-gesellschaft VI Part 2, 205-374 (1934).
6. J. Behringer, in *Raman Spectroscopy*, H. Szymanski (Ed.) (Plenum, 1967).
7. D. Gill, R. G. Kilponen, and L. Rimai, Nature 227, 743 (1971).
8. L. Rimai, D. Gill, and J. P. Parsons, J. Am. Chem. Soc. 93, 1353 (1971).
9. T. V. Long, T. M. Loehr, J. R. Alkins, and W. Lovenberg, J. Am. Chem. Soc. 92, 1809 (1971).
10. T. C. Strekas and T. G. Spiro, Biochim. Biophys. Acta 263, 830 (1972).
11. P. R. Reed, Spex Instruments Co., Metuchen, N.J. (private communication, 1972).
12. H. Brunner, A. Mayer, and H. Sussner, J. Mol. Biol. 70, 153 (1972).
13. T. G. Spiro and T. C. Strekas, Biochem. Biophys. Acta 278, 188 (1972).
14. H. Brunner and H. Sussner (in press).
15. W. Kiefer and H. J. Bernstein, Mol. Phys. 23, 835 (1972).
16. A. C. Albrecht and M. C. Hutley, J. Chem. Phys. 55, 4438 (1971).
17. H. W. Schrotter, Naturwiss. 23, 607 (1967).
18. R. S. Tobias, J. Chem. Ed. 44, 7 (1967).
19. F. A. Cotton, *Chemical Applications of Group Theory* (Interscience, New York, 1967).
20. J. R. Ferraro, *Introductory Group Theory* (Plenum, New York, 1967).
21. H. H. Jaffe and M. Orchin, *Symmetry in Chemistry* (Wiley, New York, 1967).
22. Y. Takahashi, H. Tadokoro, and Y. Chatani, J. Macromol. Sci. B2, 361 (1968).
23. E. H. Merz, G. C. Claver, and M. J. Baer, J. Polymer Sci. 22, 325 (1956).

24. R. C. Lord and G. J. Thomas, Jr., Spectrochim. Acta 23A, 2551 (1967).
25. W. Luttke, A. de Meijere, H. Wolff, D. H. Ludwig, and H. W. Schrotter, Ang. Chem. (Int'l. Ed.) 5, 123 (1966).
26. U. Bastiansen and A. de Meijere, Ang. Chem. 5, 124 (1972); T. Kasuya and T. Kojima, J. Phys. (Japan) 18, 364 (1963).
27. L. S. Bartell and H. K. Higginbothom, Inorg. Chem. 4, 1346 (1965).
28. K. K. Chiu, H. H. Huang, and L.H.L. Chia, J. Chem. Soc. Perkin II, 286 (1972).
29. C. J. Brown, Acta Cryst. 1, 97 (1964).
30. H. Ziffer and I. W. Levin, J. Org. Chem. 34, 4056 (1969).
31. K. W. F. Kohlrausch, *Ramanspektren* (Becker and Erler, Leipzig, 1943).
32. J. H. Hibben, *The Raman Effect and Its Chemical Applications* (Reinhold, New York, 1939).
33. L. J. Bellamy, *The Infra-red Spectra of Complex Molecules* (Wiley, New York, 1958).
34. K. Nakanishi, *Practical Infrared Absorption Spectroscopy* (Holden-Day, 1962).
35. S. K. Freeman and P. R. Reed, Paper presented at the Pittsburgh Conference, Cleveland, Ohio, March 1973.
36. T. G. Spiro and T. C. Strekas, Proc. Nat. Acad. Sci. U.S. 69, No. 9, 2622 (1972).
37. N. M. Amer, Y. R. Shen, and H. Rosen, Phys. Rev. Lett. 24, 718 (1970).
38. B. J. Bulkin and F. T. Prochaska, J. Chem. Phys. 54, 635 (1971).
39. N. M. Amer, and Y. R. Shen, J. Chem. Phys. 56, 2654 (1972).
40. C. H. Wang and A. L. Leu, J. Am. Chem. Soc. 94, 8605 (1972).
41. A. Saupe, Angew. Chem., Int'l. Ed. (Engl.) 7, 97 (1968).
42. L. D. Barron and A. D. Buckingham, Mol. Phys. 20, 1111 (1971).
43. L. D. Barron, J. Chem. Soc. A, 2899 (1971).
44. Y.-N. Chiu, J. Chem. Phys. 52, 4950 (1970).
45. Y.-N. Chiu, J. Chem. Phys. 52, 3641 (1970).
46. L. D. Barron, J. Chem. Soc. 2399 (1971A).
47. L. D. Barron, M. P. Bogaard, and A. D. Buckingham, J. Am. Chem. Soc. 95, 603 (1973).

48. L. D. Barron, M. P. Bogaard, and A. D. Buckingham, Nature 241, 113 (1973).
49. B. Bosnich, M. Moskovitz, and G. A. Ozin, J. Am. Chem. Soc. 94, 4750 (1972).
50. M. Diem, J. L. Fry, and D. F. Burow, J. Am. Chem. Soc. 95, 253 (1973).
51. L. D. Barron and A. D. Buckingham, Chem. Comm. (in press).

Chapter 2

INSTRUMENTATION AND SAMPLE HANDLING

2.1 INSTRUMENTATION*

2.1.1 Laser Sources

The first Raman spectrum of an organic compound was observed using the sun as a source, a telescope as a receiver, and the human eye as a detector (1). Many problems confronted the investigator in pre-laser Raman days. The low-energy light sources required at least 10 ml of a carefully distilled, clear, colorless liquid. Even then, chances of obtaining a spectrum were no better than even, and the spectra usually were poor and difficult to interpret. Photographic exposure times as high as several days were necessary to obtain a line spectrum, and fluorescence, caused by small quantities of impurities, often swamped the extremely weak Raman bands. Furthermore, sample decomposition was not infrequent with the commonly employed mercury arc source. The best to evolve, the so-called Toronto arc, generated a great deal of extraneous heat when operated under conditions necessary to achieve an intensity level sufficient to obtain acceptable Raman spectra.

Light produced from a nonlaser source is emitted from many points independently so that there is no phase relationship, no directionality, and no polarization of the radiation.

*A laser Raman schematic is shown in Figure 2.1.

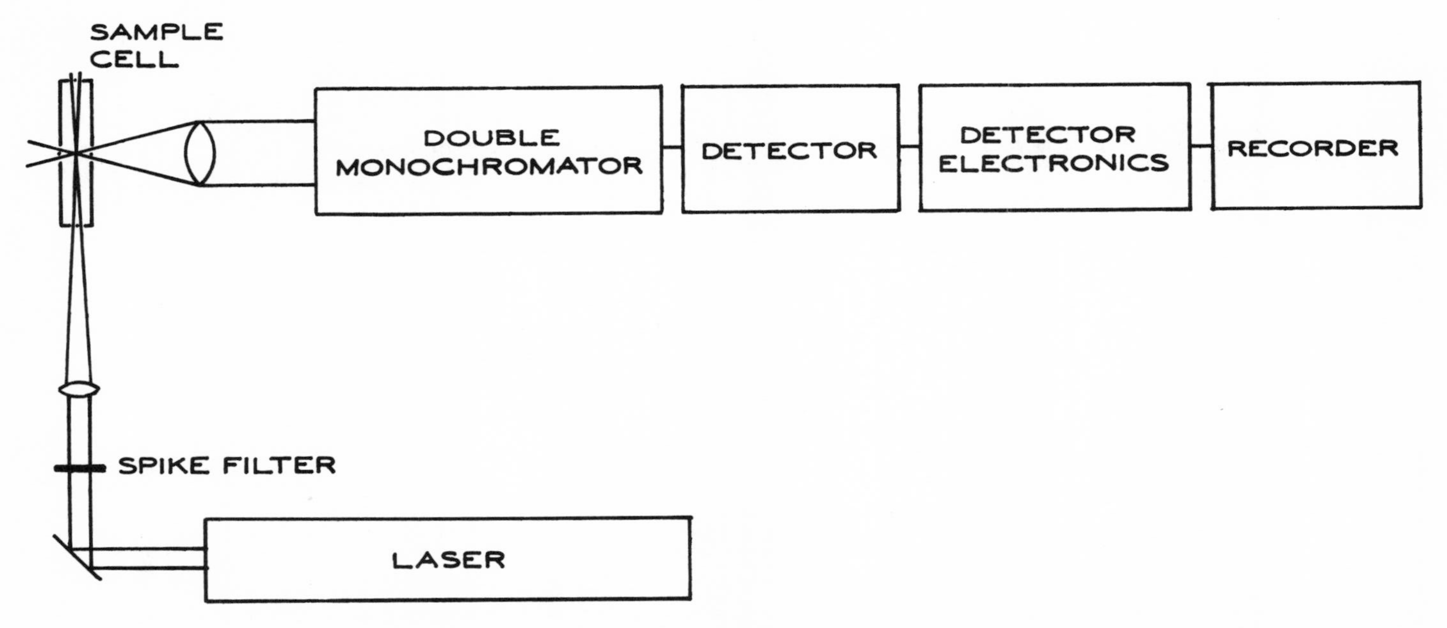

Figure 2.1. Schematic of a laser Raman spectrometer.

While the overall light intensity of such a noncoherent source may be high, the out-of-phase transverse waves produced cannot be combined to improve the brightness (energy per unit area). The laser supplied chemists and spectroscopists with a nearly ideal light source, essentially emancipating Raman spectroscopy from the disadvantages accompanying mercury arc excitation. Some of the advantages are: (1) most fluorescence problems are eliminated, (2) the collimated nature of laser energy allows focusing for excitation of extremely small volumes, and (3) the laser's brightness and nearly complete linear polarization simplify the measurement of depolarization values.

Power outputs and related data of various commercially available lasers appear in Table 2.1 (2). The signal intensity values, as photon counts per second, are based on the trisubstituted double bond stretching mode of linalool at 1685 cm^{-1} (Figure 2.2) (3). This band was selected because it represents an "average" intensity line. (By comparison, the strong C-Cl stretching vibration of CCl_4 at 459 cm^{-1} has a peak signal approximately ten times more intense.) A much greater disparity is observed between C=C band intensities employing 400 mW blue (4880 nm) and 50 mW red (632.8 nm) excitations when instrumental conditions are adjusted to yield the same peak-to-peak background noise (0.1%) and the same recording time (4 min). An increase occurs in signal strength from eightfold, based on laser power alone, to approximately fortyfold in favor of the blue radiation. This effect is due to the fact that short wavelength light has a greater molecular scattering efficiency than long wavelength light--proportional to the fourth power of the frequency. Until the latter part of 1971, another factor placed red lasers at a decided disadvantage: photomultiplier detectors incorporated in Raman instruments were more sensitive to short wavelength light. Consequently, in addition to the large overall intensity difference between spectra recorded with blue and red lasers, as well as the differences in their useable power levels, the relative intensities of bands were dissimilar for blue and red radiation. This difference in band intensities made it difficult to conveniently compare spectra recorded with different frequency lasers.

TABLE 2.1. Comparisons of Some Commercially Available Lasers

Type of Laser	Excitation Wavelength (nm)	Average Power at Sample (mW)	Peak Signal[a] (Intensity)	Approximate Peak-to-Peak Noise in Background (% of full scale)	Time (min) to Record Raman Spectrum 0-4000 cm^{-1} at 4-cm^{-1} Resolution
Cd-He	Cd 441.6 (violet)	20	6×10^4	0.4	10
Ion Ar or Ar-Kr	Ar 488.0 (blue)	400[b]	1.2×10^6	0.1	4
Ion Ar or Ar-Kr	Ar 514.5 (green)	400[b]	8×10^5	0.1	4
He-Ne	Ne 632.8 (red)	50	1.5×10^5	0.3	8
Ion Kr or Ar-Kr	Kr 647.1 (red)	200	5×10^5	0.1	4

[a]Counts per second. With reference to the C=C stretching vibration at 1685 cm^{-1} of linalool, using a cooled (-30°C) RCA C31034 photomultiplier tube.

[b]Maximum useable power for microsamples in capillary cells using a focused laser beam.

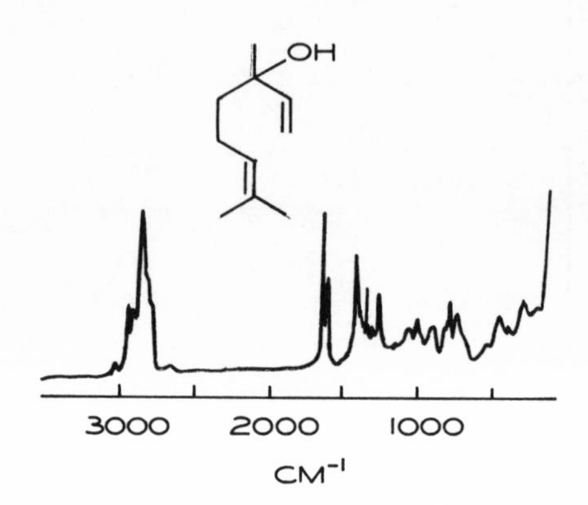

Figure 2.2. Raman spectrum of 0.8 μl linalool in a 1 mm bore borosilicate capillary, with Ar+ 488.0 nm excitation (3).

About 500 mW of blue radiation, measured at the sample, generally is the maximum useable power for milligram quantities of liquids and solids in capillary tubes. Higher energies often bring about sample decomposition. Larger amounts of liquids can be subjected to higher power because the heat generated at the focal point of the condensed laser beam is partially dissipated in the surrounding material.

It is advisable to have at least two exciting lines of 200mW minimum each, preferably blue and red. A good source is an Ar-Kr mixed gas laser emitting 200mW minimum in the blue (488.0 nm), green (514.5 nm), and red (647.1 nm) regions. Although the blue line (Ar^+) probably will be used most of the time, some organic chemicals are best examined with red radiation (Kr^+) because of fluorescence or decomposition at shorter wavelengths. Alternatively, an Ar^+ laser, capable of supplying at least 200 mW of blue and green radiation, readily interchangeable with a 50 mW minimum He-Ne laser (632.8 nm) or a 600 mW minimum Kr^+ laser (647.1 nm), make an excellent combination. With the advent of a recently designed phototube detector (see Table 2.1), red lasers are quite satisfactory in many instances.

In contrast to the previously described lasers which emit discrete lines, commercially available tunable dye lasers soon may find general use as Raman sources. The complete spectral output of an argon ion laser (457.9 to 514.5 nm) is used to "pump" an organic dye solution so that it fluoresces strongly at somewhat longer wavelengths. These dye lasers presently are tunable between 540 and 640 nm, and it is likely that the range will be increased to about 400 and 750 nm within two or three years. Wavelengths as low as 260 nm are obtained when dye solutions are pumped by pulsed lasers. Tunable lasers have found applications in resonance Raman spectroscopy, particularly for biological materials (see pp.).

2.1.2 Sampling Compartment

The laser beam is directed into the sample area by means of dielectric mirrors that maintain the polarization of the laser and reflect more than 99% of the incident light. The radiation is brought to a focus by a simple lens and the sample is accurately placed at the focus with a movable stage. Sufficient space should be available for various accessories, for example, low temperature cell, goniometer, multipass cells, and furnaces. Eye safety features must be included.

2.1.3 Monochromator

Commercial spectrometers are equipped with double or triple monochromators that effectively reject stray light. A double monochromator is satisfactory for chemical studies. Seldom does one require spectral slit widths less than 2 cm^{-1} or greater than 10 cm^{-1}. Normal scans usually are recorded at 4 cm^{-1} resolution. Raman spectrometers generally are operated at fixed instrumental slit widths, the spectral slit widths decreasing on scanning from low to high frequencies. When using a blue laser (488.0 nm Ar^+), a 4 cm^{-1} resolution at the exciting line becomes 3 cm^{-1} at 4000 cm^{-1}. It should be noted that resolutions for a typical grating IR spectrometer, operated in the survey mode, are: 4.0 cm^{-1} at 3000 cm^{-1}, 2.6 cm^{-1} at 1700 cm^{-1}, 2.0 cm^{-1} at 1000 cm^{-1}, and 8.0 cm^{-1} at 500 cm^{-1}. On the other hand, Fourier transform infrared systems present data at constant resolution over the entire spectral range.

Frequency accuracy and reproducibility of Raman bands should be ±2 cm^{-1}. Spectral scanning speed capability should range from approximately two minutes to three hours for a 4000 cm^{-1} scan. Computer compatibility is an important factor; usually the monochromator gratings are driven digitally, thus making the frequency (cm^{-1}) available as a digital signal.

2.1.4 Detector

The light leaving the exit slit of the monochromator is collected and focused on the cathode surface of a photomultiplier tube where it is converted to an electrical signal. Photomultipliers used in Raman instrumentation generally contain either GaAs (RCA 31034) or ERMA (ITT FW 130) photocathodes. In photon counting applications, the performance of the GaAs and ERMA types is directly related to the quantum efficiency. At 700 and 800 nm, tubes with the GaAs cathode photomultiplier give a threefold and a tenfold increase in signal, respectively, compared to the ERMA. The author has found ITT-FW130 photomultiplier tubes to be adequate, but a selected RCA-C31034 tube (see p.41) is preferred.

2.1.5 Detector Electronics

A combination of photon counting (pulse height analysis) and D.C. amplification systems is best able to detect weak and strong signals, respectively.

2.1.6 Recorder

A coupled flat bed or strip chart recorder can be used to display Raman spectra. An IR compatible format is helpful in facilitating IR and Raman comparisons.

2.2 FLUORESCENCE AND ITS REDUCTION

Although much less serious with laser sources, fluorescence may pose a problem in recording the Raman spectrum of a commercial chemical, a reaction mixture, a natural product, a polymer, and so on. Since the quantum yield in Raman spectroscopy is less than one millionth of that in fluorescence, traces of fluorescing impurities that do not interfere with other spectral disciplines can completely mask a Raman spectrum. Fluorescence usually can be lessened appreciably or eliminated by means of the following methods.

2.2.1 "Drench-Quench" Method

Fluorescence often is greatly reduced by exposing the sample to 200-500mW blue laser radiation. Significant decay may be attained within a few minutes, but several hours sometimes is required. In fact, some refractory materials must remain in the light path for one or two days before a Raman spectrum can be obtained. If decomposition occurs, a red laser source should be utilized.

2.2.2 Different Laser Frequencies

When the amount of fluorescence depends on the excitation frequency, a multi-line laser may be used to reduce it. Also, since fluorescence spectra sometime peak at different spectral frequencies for different excitations, a Raman spectrum can be obtained selecting "windows" for each source. This approach is employed to generate the Raman spectrum of a material which fluoresces naturally.

2.2.3 Addition of a Quenching Agent

Suppression of fluorescence may be accomplished by addition of less than 1% nitrobenzene to a liquid sample, or 1% dinitrobenzene to solids during recrystallization. This method suffers from the disadvantage that the strong NO_2 band near 1350 cm^{-1} appears in the spectrum (4).

2.2.4 Distillation, Recrystallization, Sublimation, and Extraction

Sample purification by these standard micro and semimicro techniques often can diminish the fluorescent effect. At least a several milligram sample should be on hand. Laser-induced fluorescence has been observed in cellulose from a variety of sources, including chemical cotton, filter paper, and microcrystalline cellulose. Transition metal ions are responsible for this phenomenon, and treatment of the cellulosic material with CO in either organic or aqueous media can reduce the level of fluorescence to approximately one-half that of the untreated sample (5). The same types of fluorescing impurities appear to be present in mono-, di-, and polysaccharides (5).

2.2.5 Filtration

The author has removed fluorescing impurities by filtering a solution of the material through a disposable pipette column containing neutral alumina and activated carbon (Figure 2.3). However, organic peroxides may decompose while passing through the column. Millipore filters also are helpful in reducing fluorescence. The solvent, which obviously must be free of fluorescence, may be removed under a stream of nitrogen at ambient temperature.

2.2.6 Gas Chromatography and Thin Layer Chromatography

Gas chromatography probably is the best method for removing fluorescing impurities. Some compounds must be passed through two columns of different polarities in order to achieve this goal. Injecting a sample directly onto the column and bypassing the detector minimizes thermal and metal catalyzed decomposition that might produce additional fluorescing impurities. Thin layer chromatography with nonfluorescent substrates and solvents is a good technique for materials of low volatility.

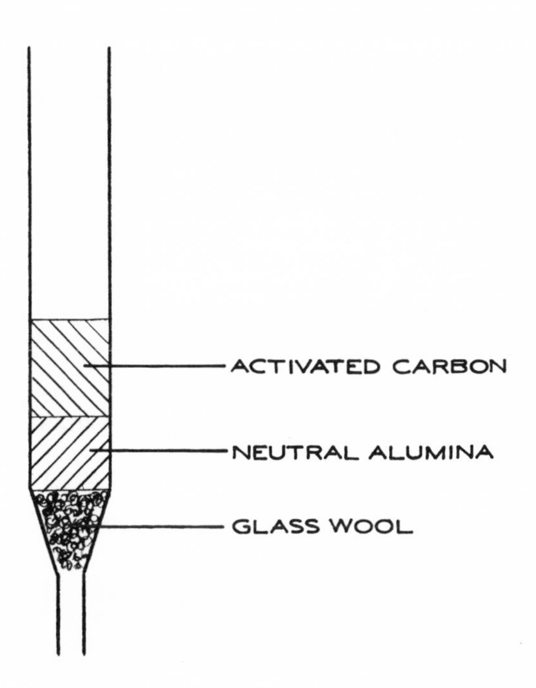

Figure 2.3. "Disposable" pipette column for purification of materials in solution to reduce fluorescence.

2.2.7 Instrumental Modifications

Derivative Spectroscopy

This technique, proposed for Raman spectroscopy, is a variant of the dual wavelength approach employed in absorption spectroscopy when, for any one of a number of reasons, a suitable reference is not available. In derivative spectroscopy a "zero" base line is produced in the presence of a signal of slowly changing intensity. Consequently, the intensity of the broad fluorescence emissions should be decreased relative to the much narrower Raman bands.

Time Discrimination

Since the Raman effect is essentially instantaneous (about 10^{-13} sec), but fluorescence requires a finite rise and decay time (about 10^{-8} sec), the principle of time discrimination to eliminate fluorescence theoretically can be applied to any sample. Therefore, a Raman scatterer illuminated by an ultrashort light pulse of 10^{-13}-10^{-14} sec will give rise to a Raman pulse approximately as wide as the exciting pulse and a fluorescence one spread over a 10^3-10^5 longer time period. A prototype optical system employing a mode-locked laser has been developed to switch beams on and off at a rate comparable to the laser pulse width, introducing a discrimination factor of this order of magnitude against fluorescence energy (6). This device may afford a practical solution to the problem of fluorescence in Raman spectroscopy.

2.3 COLORED SAMPLES

Until recently, the choice of laser frequency was restricted in the case of a colored material. When a high intensity exciting line is used which falls entirely within the visible absorption band of a sample, rapid absorption of the radiation may cause local overheating and decomposition. For example, a yellow, orange, or red material may be destroyed at the focus of green (514.5 nm) or blue (488.0 nm) excitation. Raman spectra can be obtained on highly absorbing liquids and solution by rotating the liquid cell at about 3000 rpm (7). A tenfold gain in signal to noise results. The relative motion between laser focus and sample surface avoids local overheating and decomposition. Solids are amenable to this rotating technique by pressing them into ring-shaped pellets (8,9) (Figure 2.4). The method differs from preparation of alkali halide pellets for IR spectral recording in that the pellet need not be mixed with an alkali halide nor fused into a transparent pellet under high pressure. For smaller samples (about 10-mg minimum), a layer of KBr is pressed into the ring and a thin layer of sample sprinkled and pressed on the surface of the KBr.

It should be noted that this technique does not decrease fluorescence, and any colored sample may give rise to a resonance Raman effect (10) (see p.10). The resonance Raman spectrum of β-carotene (11) displayed in Figure 2.5 was obtained by placing the surface of a raw carrot at the focal point of an argon ion laser.

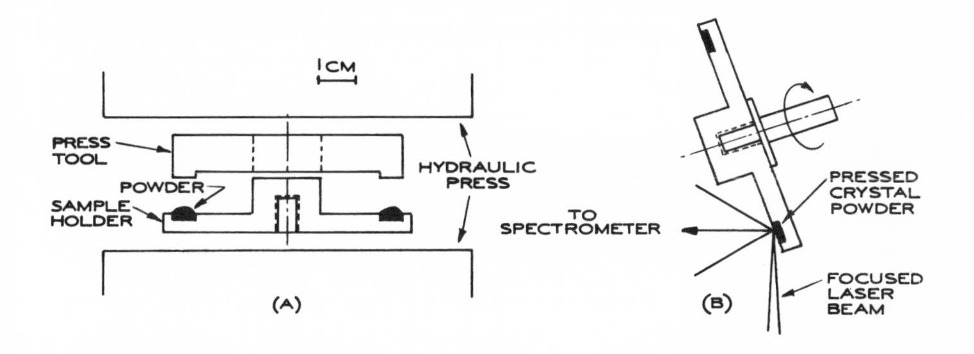

Figure 2.4. (a) Sample holder with powder in ring in hydraulic press. (b) Rotating accessory in position in spectrometer (8,9).

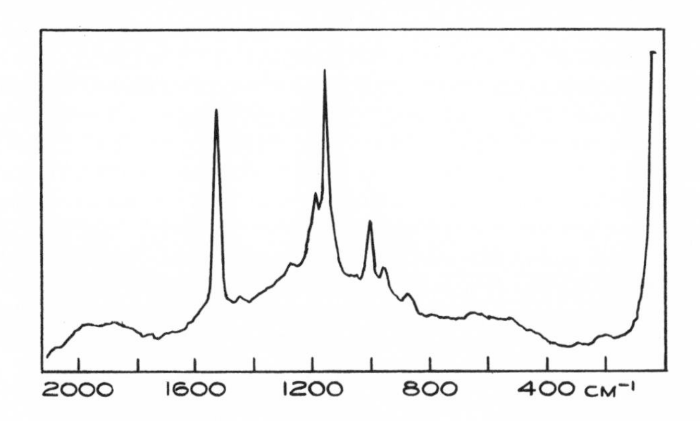

Figure 2.5. Raman spectrum of β-carotene (obtained by placing a raw carrot at the focal point of an Ar^+ ion laser), 488 nm excitation (11).

2.4 SAMPLE HANDLING TECHNIQUES

2.4.1 Introduction

When Raman, NMR, IR, mass, and UV data are required and only a small quantity of material is available, a Raman spectrum should be obtained first. Mass spectroscopy is a destructive technique and it is customary for the sample to be dissolved in a solvent for NMR and UV. A liquid must be presented to the IR spectrometer spread between alkali halide plates or contained in a micro cell; solids have to be dissolved in a solvent or pressed into disks with potassium bromide. On the other hand, a Raman spectrum can be gained directly on the neat liquid or solid sample in a capillary tube. Subsequently, a small portion of the material may be withdrawn for mass spectral study. An IR recording can be gained on the remaining sample which is then transferred with an appropriate solvent from the salt plates or sealed cell to an NMR tube or a UV cuvette.

A variety of Raman cells are available for the examination of solids, liquids, and gases. In the author's view, there is no difference between a 2-g and a 2-mg quantity of solid or liquid with regard to convenience in presenting it to the spectrometer, or in the quality of the resulting spectrum. For this reason the following discussion on liquids and solids will be concerned mainly with milligram and submilligram quantities. The beam diameter of laser wavelengths employed in Raman spectroscopy is approximately 2 mm and it is reduced to about 20 μ after focusing. The focused beam volume is approximately 10^{-2} nl (about 10 ng), which represents the minimum sample volume expected to yield a useful spectrum (2).

2.4.2 Liquids

Capillary Containers

A melting point capillary tube is an excellent container for either liquid or solid material (12). After partially filling with sample (a 3-mm continuous liquid column is approximately 2 μl), the open-ended tube is placed in the sampling compartment of the instrument and properly positioned by means of a movable stage, and so forth (Figure 2.6).

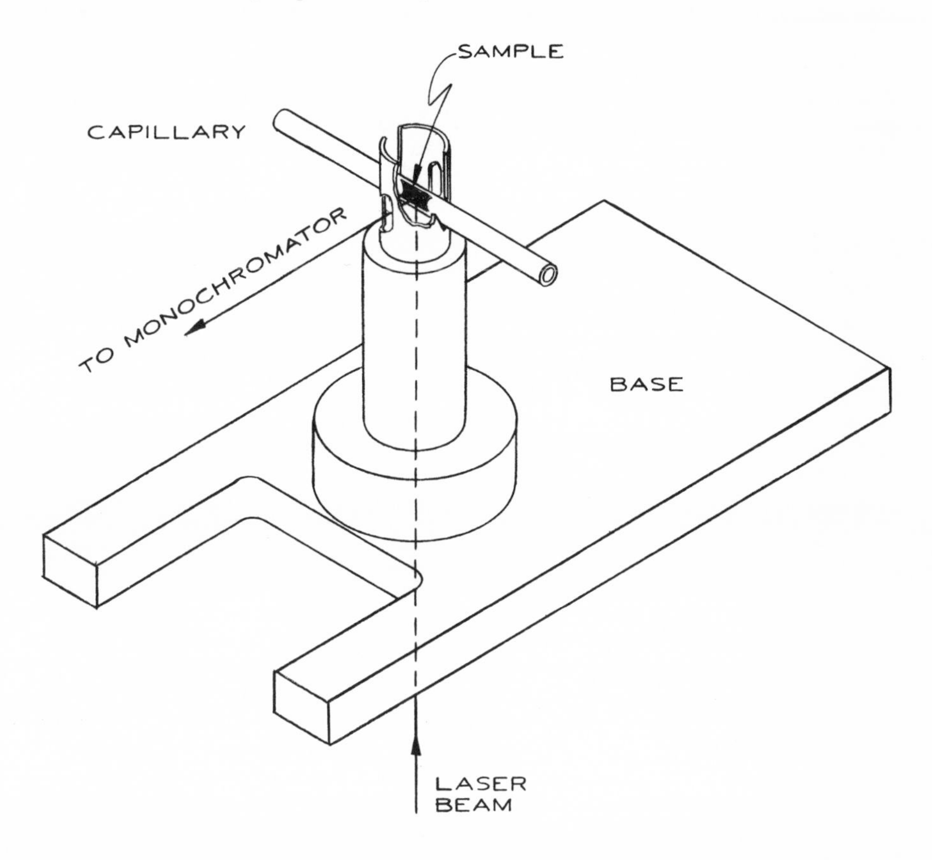

Figure 2.6. Sample holder for capillary tubes. (Transverse illumination, transverse viewing.)

For low boiling materials, the capillary should be sealed to prevent evaporation. Once a capillary is properly aligned at the focused region of the laser it can be readily substituted by another sample in a capillary of approximately the same diameter after a spectrum has been recorded. Survey curves can be performed in approximately two minutes employing post-1970 commercial laser Raman spectrometers. Flat bottom capillaries or those with a hemispherical lens (13) require a different viewing geometry: axial excitation, transverse viewing. These cells are not as convenient to manipulate routinely.

Smaller quantities can be handled in narrower bore capillaries. For example, Figure 2.7 shows a spectrum recorded in three minutes on 2-nl indene in an 0.1-mm i.d. tube, drawn from a standard melting-point capillary (2).

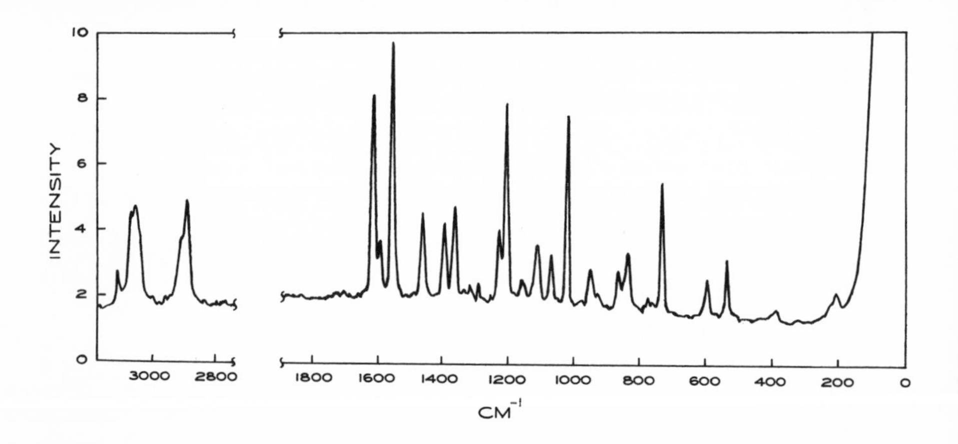

Figure 2.7. Raman spectrum of 2-nl indene in an 0.1-mm capillary, 3-min. scan, 250 mW, 514.5-nm excitation (8 cm^{-1} resolution) (2). [Courtesy of Mikrochimica Acta]

Ordinarily, a respectable spectrum (4 cm^{-1} resolution) can be produced with nanoliter quantities in approximately fifteen minutes using a 200 mW blue laser. Highly polar compounds such as water and methanol are poor Raman scatterers and therefore are excellent solvents. A spectrum of 2 nl of a 20% aqueous solution of cysteine hydrochloride is shown in Figure 2.8 (2).

Increasing the cell path length above 1 mm does not result in a more intense signal because the useable slit height of most laser Raman instruments is approximately 10 mm (about a tenfold magnification of the laser-illuminated region in the sample is imaged on the entrance slit of the monochromator). However, for capillary diameters less than 1 mm, signal strength is inversely proportional to path length. Consequently, a tenfold difference in spectral intensity will be observed when comparing recordings of a sample in 1.0- and 0.1-mm capillary cells.

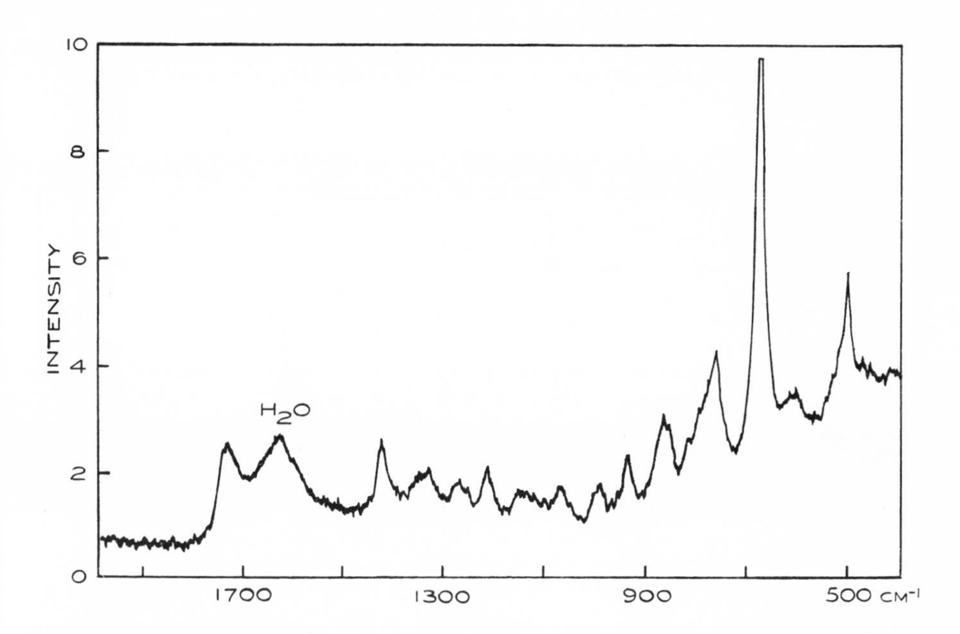

Figure 2.8. Raman spectrum of 2-nl cysteine·HCl (20% aqueous solution) in an 0.1-mm i.d. capillary, 300 mW, 514.5-nm excitation (4 cm^{-1} resolution) (2). [Courtesy of Mikrochimica Acta]

At least a tenfold signal amplification is required to restore this energy loss. This, of course, adversely affects the signal-to-noise ratio, necessitating spectral recording at slower speeds or with wider slits.

Preliminary evidence indicates that long low-loss liquid-core optical fibers can be exploited as liquid cells (14). Intense Raman radiation has been obtained from benzene and tetrachloroethylene by passing the focused beam of a 488.0 nm Ar^+ laser through 10-to-25-m filled hollow fused-quartz optical fibers having core diameters of about 75 μm. The principal advantages gained from the optical fiber technique are that high

Raman intensities are obtained from the long paths when the losses are low, that it is possible to use relatively small sample volumes, and that a large proportion of the Raman radiation is collected. On the other hand, the limitations of this method are that accurate depolarization measurements are precluded and that fused-quartz fibers limit the liquids that can be examined to those having refractive indices above 1.463 at 488.0 mm. However, with the availability of other fibers with lower indices of refraction, such as Teflon (η = 1.35), a wider range of pure liquids as well as aqueous solutions could be investigated.

Gas Chromatographic Fractions. Routinely, sample isolation for Raman examination may be achieved by trapping into inexpensive, sturdy, 0.7-mm i.d. borosilicate glass melting-point capillaries (12). One end is then heat sealed and the material centrifuged to the closed end. Depending on sample size, a 0.1-to-0.3-mm-bore capillary is threaded into the 0.7-mm i.d. tube; capillary action effects the transfer. Alternatively, the opposite end section of the sealed melting-point tube is drawn to approximately 0.2-mm i.d. and sealed; then the sample is either centrifuged to the drawn end or gently heated and condensed into the constricted section by being cooled with a thin metallic strip or with metal forceps. Isolated solids are concentrated to a narrow band by being heated while the capillary is maintained in a vertical position.

Capillaries as narrow as 0.2 mm i.d. are compatible with most gas chromatographs for trapping both liquids and solids (12,15,16). A flow reduction of only 15% occurs when a 0.2-mm-bore glass capillary is inserted into a gas chromatograph equipped with a 1/8-in i.d. column operated at a helium flow of approximately 50 ml/min. A negligible decrease is observed when a 0.2-mm-bore capillary is used for trapping fractions from a support coated open tubular (SCOT) column. For low boiling materials, capillaries can be cooled with dry ice or with a cold stream of air or nitrogen in order to prevent large losses of eluants. Melting-point capillaries can be used with sections drawn to 0.1-0.4 mm i.d. (17). They are cooled by a small coil of copper wire wound round the capillary (downstream from the constricted section), the ends of which are immersed in liquid nitrogen. If the sample collects as droplets they can be centrifuged to the sealed end of the constricted portion of the capillary. Fogs formed on the glass surface usually can be converted to a droplet by sealing both ends of the capillary and heating gently while holding the capillary with a small pair of forceps. The sample forms into a bead where the capillary is in contact with the forceps. In order to prevent the material from redispersing, the sample should be centrifuged immediately to the end of the constricted portion, the larger part cut off, and the resulting small bore capillary sealed.

Thin Layer Chromatographic Fraction. It is not generally practical to record the spectrum of a compound directly from a TLC plate because of the relatively small number of molecules intercepting the laser beam. However, in some instances signal averaging may be of value. A Raman spectrum often can be generated on a sample obtained by solvent extraction of a desired TLC section, evaporating to a few microliters, followed by transferral to a capillary tube where further volume reduction can be effected. Substrates for TLC must be free from fluorescing additives.

Noncapillary Containers

Sealed ampules, reagent vessels, and so forth, can be placed directly in the laser light path, and spectra of materials contained in red or brown bottles can be obtained with red radiation. A commercially available multipass cell of about a 2-ml capacity may be used for dilute solutions or for determining the presence of minor impurities. The laser beam traverses the cell about forty times, yielding an approximately tenfold gain in signal intensity.

Rotating Raman Cell for Difference Spectroscopy

The rotating Raman sample technique (see p. 49) developed to study highly absorbing materials can be used for solutions (18). This method is simpler than ordinary Raman difference spectroscopy (19), where the laser radiation is split into two beams which then are focused into two different sample cells. The rotating Raman cell for samples in solution is divided into two equal sections (Figure 2.9), one containing the solution, the other the solvent.

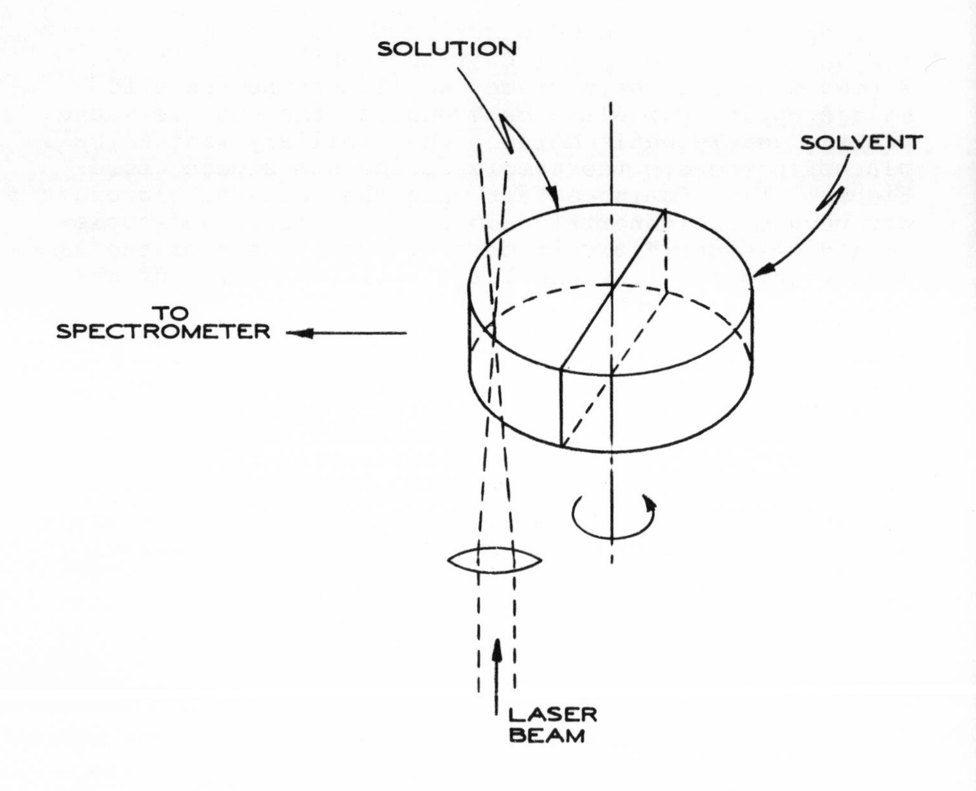

Figure 2.9. Schematic diagram of the rotating cell for Raman difference spectroscopy (18).

Rotation of the cell exposes the laser beam alternately to the solution and the solvent. Thus the spectrometer and detector alternately receive signals from the two different liquids. Signal subtraction can be accomplished electronically and a Raman difference spectrum generated. This approach also could be used to detect small quantities of impurities in liquids by placing the pure liquid and the sample in the rotating cell.

2.4.3 Solids

General

A powder or fine crystal may be transferred into a melting-point capillary by thrusting it into the sample vial or reagent bottle. The capillary is then placed in the spectrometer sampling compartment (see Figure 2.6). Coarse solids weighing a few milligrams can be ground to a powder in a small mortar and transferred to a capillary by tamping. Smaller samples, as little as a few nanograms, are best handled by dissolving in a low boiling solvent, picking up the solution in a fine capillary, and allowing the solvent to evaporate while the cell is suspended in a vertical position (2). Recording the spectrum of an air or moisture sensitive material can be accomplished by pouring the powder into an ampule, evacuating, and sealing. Front reflection, or a 90° arrangement, [Figure 2.10(a)] affords about twice the intensity of scattered radiation, but 10-1000 times as much exciting light enters the spectrometer compared to passage of laser radiation through the sample as in forward scattering (20) [Figure 2.10(b), 0° geometry].

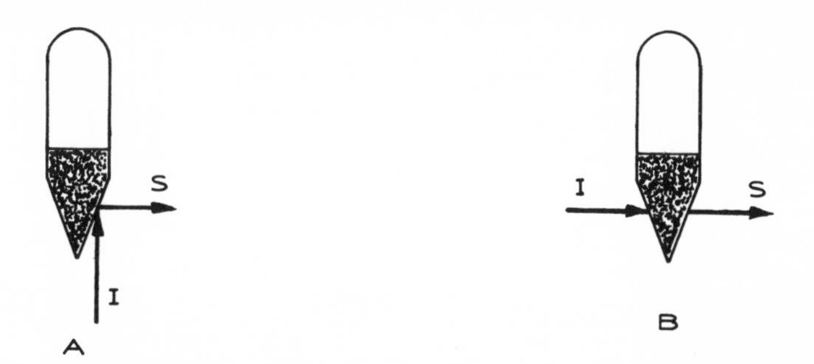

Figure 2.10. Examination of crystalline powders in ampules. (a) Front scattering; (b) forward scattering. I = incident light; S = scattered light (20).

At times this exciting radiation produces stray light which interferes with the Raman spectrum. Intact samples, such as a raw carrot (see p.49) or an aspirin tablet (Figure 2.11) (21) can be examined directly by placing the materials at the focal point of the laser beam.

A prototype Raman instrument appears to be capable of generating spectra on extremely small samples in the femtogram (10^{-15}g) range (22).

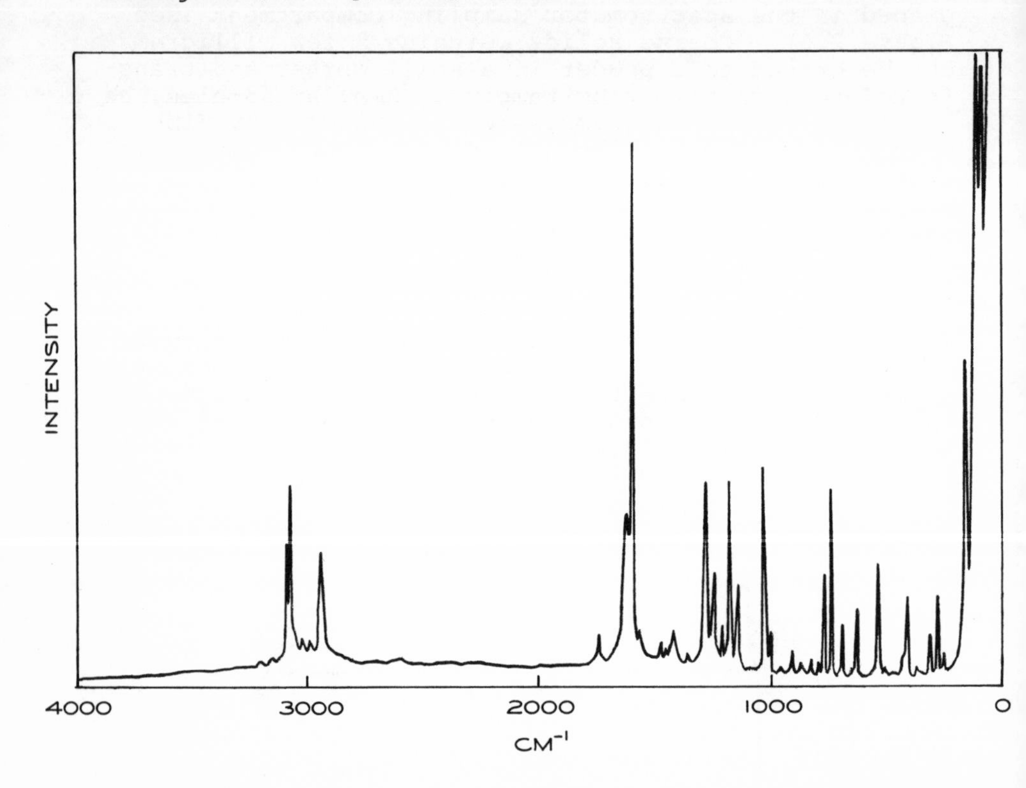

Figure 2.11. Raman spectrum of an aspirin tablet (21).

The system consists of an SEC vidicon detector scanned and read out by a small dedicated computer, an echelle spectrometer with crossed dispersion, and a nearly square entrance slit. The sample optics are based on a special microscope with very long tube and focal lengths whose objective is concentrically surrounded by an ellipsoidal collecting mirror operating through an annular cutout on a thermoelectrically cooled stage.

Single Crystals

Although there is a large body of Raman literature devoted to single crystal studies, little information in it is of practical value for the organic chemist. Raman spectra of single crystals are easy to record and useful data can be gained for virtually any size or shape of crystal. Individual crystals ordinarily are best handled by cementing them to a goniometer, a device employed in X-ray diffraction studies that allows simple traversal in the x, y, and z directions. Special adhesives are available, but modeling clay is satisfactory for sturdy crystals. The spectrum of a few micrograms of a pharmaceutical product is shown in Figure 2.12 (2). IR spectroscopy, being an absorption technique, requires crystals of about 10-μm light path, and it is virtually impossible to prepare single crystals of this thickness.

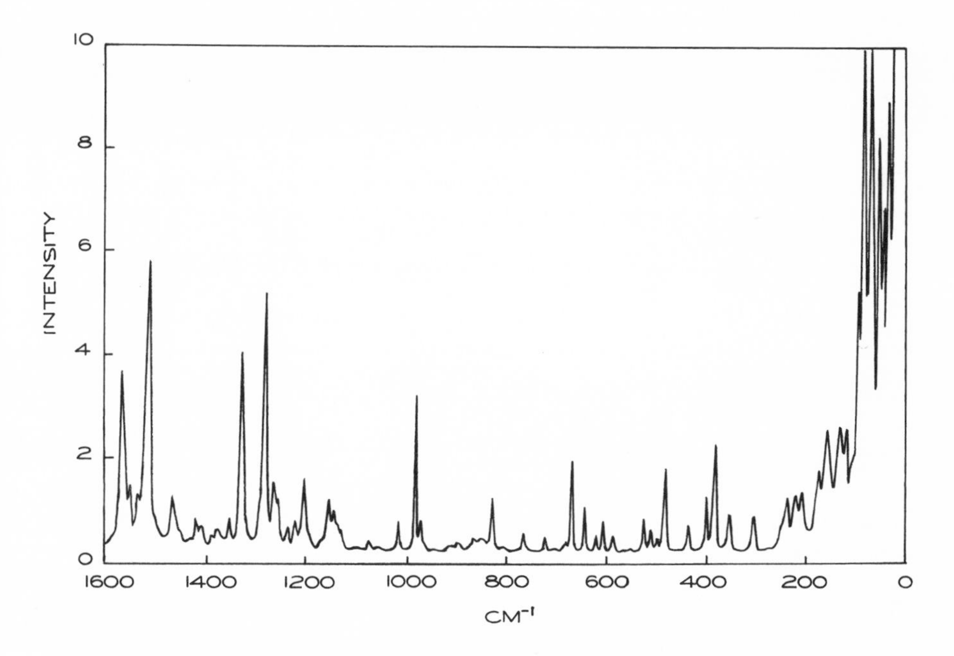

Figure 2.12. Raman spectrum of a single crystal of chlordiazepoxide·HCL (about 3 μg), 200 mW, 568.2-nm excitation (2 cm^{-1} resolution) (2). [Courtesy of Mikrochimica Acta]

Steroids

Steroids usually require stringent purifications to remove fluorescence (4). The dissolved steroid is shaken with activated charcoal, heated to 50° on a water bath, and passed through an approximately 1-cm column containing aluminum oxide. Two or three grams of aluminum oxide are required per gram of steroid. The dissolved, purified steroid is exposed to laser radiation in the spectrometer in order to further reduce fluorescence. The solvent is allowed to evaporate slowly in order to crystallize the material. Fluorescence may be observed for some crystalline steroids even when the dissolved, purified substances are fluorescence free. Steroids possessing conjugated carbonyl groups are seldom completely rid of this phenomenon.

Raman spectra of steroids have been obtained by directly irradiating pressed pellets (front scattering) (4). Figure 2.13 portrays the intensity of Raman radiation as a function of the pellet thickness with different scattering modules, r, that take into account the reflection, refraction, and scattering of the exciting light.

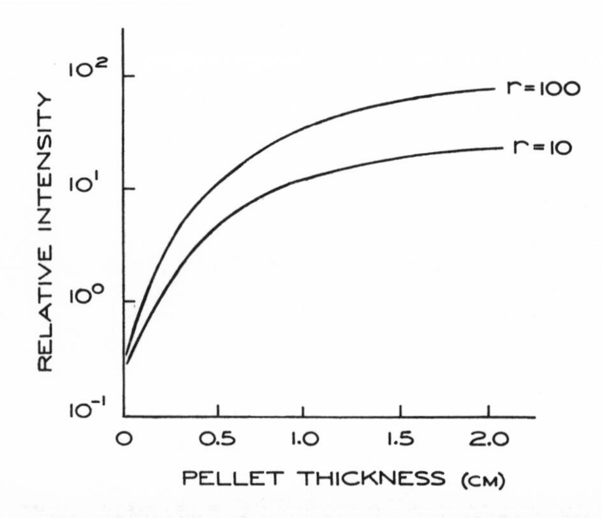

Figure 2.13. Dependence of Raman intensity on thickness of steroid pellet (4).

The scattering module is inversely proportional to the particle size of the granules. The intensity of Raman scattering reaches a maximum with optimum layer thickness of the pressed tablet. Coarse crystalline substances exhibit larger scattering than fine ones, but a reduction of "r" for a particular granular size is possible when the material is compressed into a tablet. An advantage of using a pellet over a crystalline powder is that the sample thickness can be regulated easily in pellet preparation. Steroid tablets of 13-mm diameter may be prepared under five tons pressure, employing between 300 and 600 mg of material for very fine to very coarse crystals. If the spectrum indicates that a disk is too thick, it can be scraped to proper size with a razor blade. Sometimes the absorption of laser light by the pellet requires a front surface scattering approach, where it is oriented about 60° to the incident beam.

Synthetic Polymers

Good spectra can be recorded from most homopolymers and even some copolymers (23). Laser energies as high as one watt may have to be used. In contrast to IR, samples do not have to transmit light. Expectedly, some polymers present difficulties if they are colored or contain dark, absorbing fillers or fluorescing impurities. For these problems, in situ examination is not possible and the polymers must be separated by standard chemical methods. Some polymers develop fluorescence on aging, but polymers themselves do not appear to fluoresce. Fillers of glass, clay, and silica are poor scatterers so spectra usually can be obtained without removing them. On the other hand, since fillers are strong IR absorbers, IR spectra cannot be recorded on intact specimens. Carbon black fillers can prevent the generation of Raman spectra because the rise in temperature caused by laser light absorption leads to sample decomposition. Raman spectroscopy is of considerable advantage over IR when very small samples of a coating, a finish, or a fiber are available, since it is necessary only to fill the approximately 20 μm focused beam to obtain an acceptable Raman spectrum. Generating an IR spectrum of a polymer is often a difficult task, particularly when the sample cannot be made into a thin film by compression molding due to infusibility or cannot be cast from a solvent owing to insolubility. Moreover, if the particle size is too large, good KBr pellets cannot be made. Foams pose a difficult problem for the IR spectroscopist because of their very low density. A Raman spectrum may be conveniently obtained on a piece of sample placed at the focus of the laser beam. Sample pretreatment, such as melting or dissolving a plastic, is undesirable if the thermal history is important for understanding the properties of the material. Unlike IR, polymers of this type, for example, thermosetting resins and cross-linked rubbers, can be studied by Raman spectroscopy without modification. Some standard Raman sampling techniques (24) are depicted in Figure 2.14.

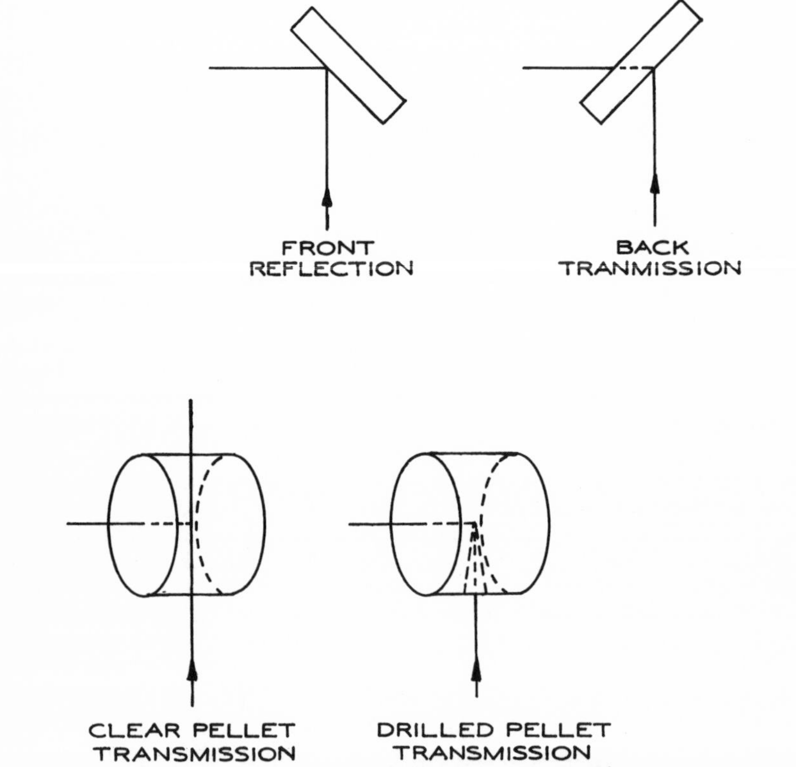

Figure 2.14. Sampling techniques for polymers (24).

Thin films can be examined conveniently by front reflection. The laser beam is brought to focus within the film, thereby attaining high intensity of illumination within a very small sample area. The 90° viewing system is generally used for examining film samples, but the 180° forward scattering arrangement [Figure 2.10(b)] may be preferable when routinely handling chips, fibers, and bulk materials.

Raman spectra can be generated on single fibers as fine as 10-50 μm (Figure 2.15) (25), a nearly impossible task for IR spectroscopy.

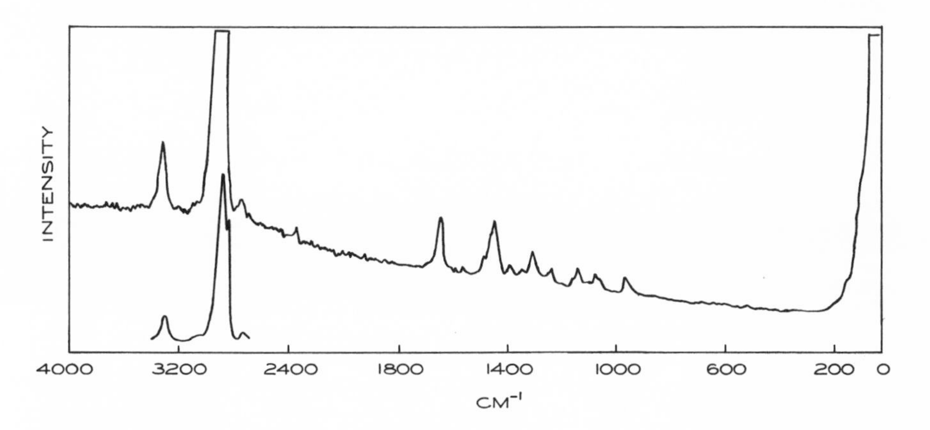

Figure 2.15. Raman spectrum of a single strand of 50 μm diameter nylon 6-6 fiber. Sample was oriented nearly parallel to slits and incoming source radiation. Thus, strand was illuminated at a grazing angle of incidence by focused laser beam (25).

Information pertaining to the orientation of a fiber can be gained from Raman measurements <u>in situ</u> as the rotational angle of the fiber is changed with respect to the laser beam. The Raman spectrum of a hard, yellow polymeric screwdriver handle (shown in Figure 2.16) was obtained with no prior sample preparation (25).

Experimentally, two basic problems that militate against the general application of Raman spectroscopy to polymer studies are the low intensity of many polymer spectra and the high background emissions other than Raman, Rayleigh, and Tyndall scattering. Fortunately, the strong Raman bands usually are sharp and thus do not obscure the weaker ones.

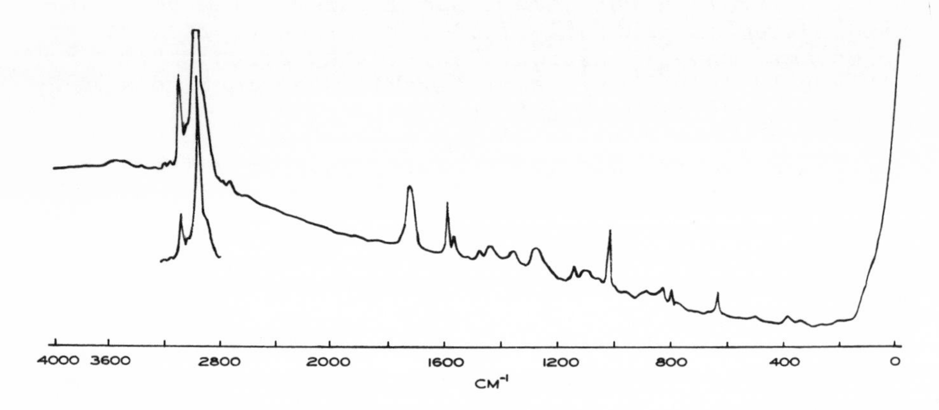

Figure 2.16. Raman spectrum of a clear, yellow, polymeric screwdriver handle (25).

The intensity of the high background generally decays when the sample is subjected to laser light (26) (Figure 2.17), but it sometimes poses a problem. Another approach, which has been reported to be quite successful (26), is the preparation of films from bulk samples. Nonoriented films are formed in a press operated under pressure at a temperature slightly higher than the melting point of the material. After releasing the pressure, the thin film is plunged into a mixture of dry carbon dioxide and methanol.

2.4.4 Gases and Rapid Scan Raman spectroscopy

A Fourier transform IR spectrometer is capable of recording spectra of gas chromatographic gaseous eluates "on-the fly" (27), but a mass spectrometer undoubtedly is the best tool to accomplish this feat (28). On the other hand, most organic chemical vapors have insufficient molecular densities to yield good Raman spectra in a reasonable time with presently available instruments, and the Fourier transform approach is not applicable under photon noise-limited conditions.

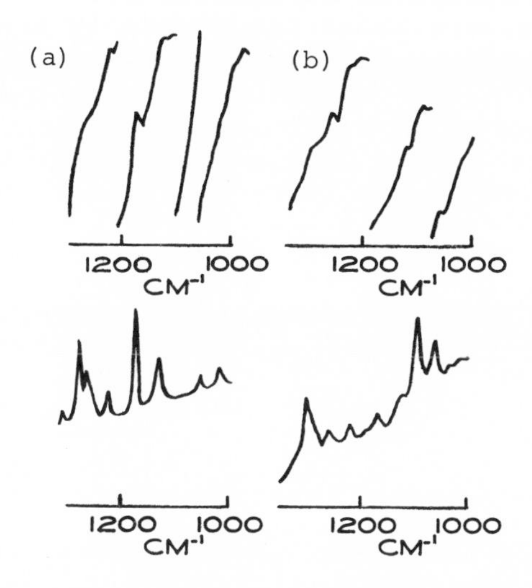

Figure 2.17. (a) Spectra of polyacrylamide. Top: spectrum after ∿1/2-h exposure; He-Ne laser at 632.8 nm. Bottom: spectrum after 10-h exposure; same conditions as (a) top. (b) Spectra of nylon 8. Top: spectrum after ∿1/2-h exposure; He-Ne laser at 632.8 nm. Bottum: spectrum after 10-h exposure; same conditions as (b) top (26).

The extreme difficulty in designing and fabricating a Michelson interferometer to meet resolution and other requirements for a Fourier transform Raman spectrometer makes it unlikely that one will be built soon. Even if such an interferometer were available, the advantage over conventional instrumentation would not be significant because only when noise originates in the detector does the signal-to-noise (S/N) ratio increase by Fourier transform multiplexing. Raman phototube detectors have low detector noise. If the noise occurs in the signal (shot, or statistical, noise), increasing the signal also brings about an increase in the noise.

The difficulty of containing a neat liquid sample in a column less than about 0.3 nl prevents recording Raman spectra of quantities as little as 0.01 nl, which is the lower limit generally dictated by laser beam diameters. A partial solution to this problem may lie in spectrally examining vapors in multipass cells with a combination of high laser power and time averaging. The latter technique has been very successfully applied in NMR (29) and shows promise in IR (27) studies. Since noise can be considered to be a random process, while the Raman signals are not, it is possible to enhance the S/N ratio by multiple scanning of a given sample. The S/N ratio is proportional to the square root of the number of scans. Successive scans can be automatically added by a computer, causing the signal intensities to increase with the number of scans and the noise to trend toward zero. The Raman region between 100 and 2000 cm^{-1} can be recorded in about one minute (4 cm^{-1} resolution) with modern instrumentation, which means that a S/N increase of about ten can be gained by repetitive scanning over a one-hour period. Obviously, similar signal enhancement is obtainable for liquids and solids. Investigations pertaining to hyper-Raman spectroscopy offer promise for determining concentrations of gases. In the hyper-Raman effect (30), the sample absorbs two photons from a pulsed laser beam, undergoes a transition to a new energy state, and reradiates a third photon, all simultaneously. Research Raman instrumentation has been developed by combining an image intensifier, an isocon camera tube, and a multichannel analyzer (31). Spectral information is collected simultaneously and extremely low signal levels can be observed by integration over several hours.

REFERENCES

1. C. V. Raman, Indian J. Phys. 2, 387 (1928).
2. S. K. Freeman, P. R. Reed, D. O. Landon, Mikrochim. Acta 288 (1972).
3. S. K. Freeman and D. O. Landon, The Spex Speaker VIII, No. 4 (1968).
4. E. Steigner, Ph.D. thesis, Westphalian Wilhelms University, Munich, Germany (1969).
5. R. H. Atalla. Chem. Comm. (in press).
6. T. Hirschfeld, Block Instrument Co. (private communication, 1973).
7. W. Kiefer and H. J. Bernstein, Appl. Spectrosc. 25, 500 (1971).

8. W. Kiefer and H. J. Bernstein, Appl. Spectrosc. 25, 609 (1971).
9. H. J. Sloane and R. B. Cook, Cary Instruments Technical Memo. No. R-71-7, Cary Instruments, Monrovia, California; Appl. Spectrosc. 26, 589 (1972).
10. J. Behringer, in *Raman Spectroscopy Theory and Practice* H. A. Szymanski (Ed.) (Plenum, New York, 1967), p. 168.
11. H. J. Sloan, Paper presented at The Pittsburgh Symposium, Cleveland, Ohio, March 1972.
12. S. K. Freeman and D. O. Landon, Anal. Chem. 41, 198 (1969).
13. G. Bailey, S. Kint, and J. R. Scherer, Anal. Chem. 39, 1040 (1967).
14. G. E. Walrafen and J. Stone, Appl. Spectrosc. 26, 585 (1972).
15. B. J. Bulkin, K. Dill, and J. J. Dannenberg, Anal. Chem. 43, 974 (1971).
16. R. P. Oertel and D. V. Myhre, Anal. Chem. 44, 1589 (1972).
17. R. O. Kagel, Dow Chemical Co. (private communication, 1972).
18. W. Kiefer, Appl. Spectrosc. 27, 253 (1973).
19. J. S. Bodenheimer, B. J. Berenblut, and G. R. Wilkinson, Chem. Phys. Letters 14, 533 (1972).
20. B. Schrader, W. Meier, E. Steigner, and F. Zöhrer, Z. Anal. Chem. 254, 257 (1971).
21. R. Obremski, Beckman Instruments Co. (1972).
22. T. Hirschfeld, J. Opt. Soc. 63, 476 (1973).
23. J. L. Koenig, Appl. Spectroscopy Reviews 4, 233 (1971).
24. W. B. Rippon, J. L. Koenig, and A. G. Watson, J. Agr. Food Chem. 19, 692 (1971).
25. H. Sloane, Appl. Spectrosc. 25, 430 (1971).
26. M. J. Gall, P. J. Hendra, D. S. Watson, and C. J. Peacock, Appl. Spectrosc. 25, 423 (1971).
27. S. K. Freeman, in *Ancillary Techniques of Gas Chromatography*, L. S. Ettre and W. H. McFadden (Eds.) (Wiley, New York, 1969).
28. F. A. J. Leemans and J. A. McCloskey, J. Am. Oil Chem. Soc. 41, 11 (1967).
29. T. C. Farrar and E. D. Becker, *Pulse and Fourier Transform NMR* (Academic, New York, 1971).
30. R. Bershon, Y. H. Pao, and H. L. Frisch, J. Chem. Phys. 45, 3184 (1966).
31. C. M. Savage and P. D. Maker, Appl. Optics 10, 965 (1971).

Chapter 3

AMINES, ALKYNES, AND NITRILES

3.1 AMINES

Of the commonly employed spectral techniques, IR probably is the best diagnostic for primary amines. Two moderately intense, sharp bands are observed at 3250-3550 cm^{-1}, originating in the asymmetric and symmetric stretching vibrations, respectively. The relative intensities expectedly are reversed when going from IR to Raman. Secondary amines display a weaker, single absorption in this region. Hydroxyl bands in IR generally are much stronger than those of amino groups, but the converse is true in the Raman effect. For example, in ethylene glycol, where the IR hydroxyl to methylene band (about 2900 cm^{-1}) intensity ratio is 1, the Raman ratio is approximately 0.1 (Table 3.1).

TABLE 3.1. Comparison Between Relative Ratios of Hydroxyl/Methylene and Amine/Methylene Groups in Raman and IR

Ratio	$HOCH_2CH_2OH$		$NH_2CH_2CH_2NH_2$	
	IR	Raman	IR	Raman
OH/CH_2	1	0.1	--	--
NH_2/CH_2	--	--	1	0.8

On the other hand, although the amine/methylene IR intensity for ethylenediamine also is 1, it is about 0.8

in the Raman spectrum (1). The intense intermolecularly hydrogen-bonded OH stretching band of 2-amino-1-butanol swamps the weaker NH_2 absorptions that are barely evident as "pips" near 3300 and 3400 cm^{-1} (Figure 3.1).

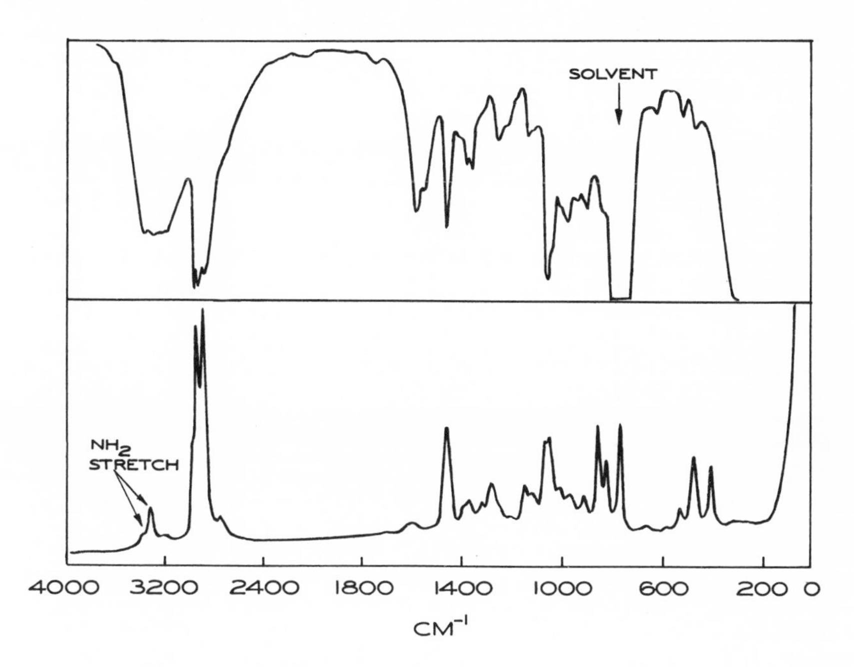

Figure 3.1. 2-Amino-1-butanol. Upper curve: IR; lower curve: Raman (1).

It is difficult therefore, to confirm the presence of the NH_2 group from the IR evidence. Inspection of the Raman spectrum clarifies the situation; the two bands near 3300 cm^{-1} correspond to the asymmetric (3380 cm^{-1}) and the symmetric (3280 cm^{-1}) NH_2 stretching modes. The OH band is not observed and the weak band at 3190 cm^{-1} is probably an overtone of the approximately 1600 cm^{-1} NH_2 deformation, intensified by Fermi resonance (2), a phenomenon frequently encountered in vibrational spectra.

3.2 ALKYNES

Of all the spectral disciplines, the Raman effect undoubtedly is the technique of choice for determining the presence of an alkyne group. The vibrations of triple bonds provide excellent group frequencies; unlike C=O stretching frequencies, the precise frequencies of the C≡C stretching modes are primarily dependent on their force constants. The C≡C stretching vibration of monosubstituted alkyl acetylenes appears in IR as a weak to moderately strong absorption near 2100 cm^{-1}, but in the Raman effect it appears as an intense band ($\rho \sim 0.1$). Since this IR band is very weak for some unconjugated alkyne acetates (Figure 3.2), the 3250 cm^{-1} C-H stretching band might be confused with an N-H stretch.

A different situation exists for disubstituted alkynes. Band intensities may be weak or zero in the IR when the substituents are similar in mass, inductive, and resonance properties. The C≡C stretching vibration is forbidden in the IR of symmetrically disubstituted acetylenes. Frequencies of monosubstituted alkynes range from 2100 to 2150 cm^{-1} and lie between 2190 and 2260 cm^{-1} for unsymmetrically disubstituted alkynes. Two bands, probably arising from Fermi resonance, are present in the Raman spectra of some dialkyl acetylenes (Figure 3.3) and alkynes of the type $RC_nC{\equiv}CC_nOR'$, where R = alkyl, n = 1 or 2, and R' = H or O=C-CH_3 (3). In a few instances, the low frequency band is doubled (Figure 3.3). The weak scatterings near 2200 cm^{-1} are ascribed to the isotopic molecules RC $\equiv^{13}$CR' (4).

Reduction of an alkyne to an alkene can be monitored conveniently and rapidly by Raman spectroscopy. Hydrogenation of acetylenes with platinum metal catalysts proceeds stepwise, the intermediate olefin being predominantly (Z), or *cis* (5).

$$RC{\equiv}CR' \xrightarrow{H_2} \overset{H}{\underset{|}{RC}}{=}\overset{H}{\underset{|}{C}}\text{-}R' \xrightarrow{H_2} RCH_2CH_2R'$$

Despite the many variables, such as amount and type of catalyst, substrate, inhibitor, temperature, pH, and agitation, good to excellent yields of either the alkene or alkane can be achieved with little difficulty. Usually only a very small amount of olefin will be hydrogenated as long as the alkyne is present because the available catalyst sites will be occupied by the more strongly adsorbed alkyne.

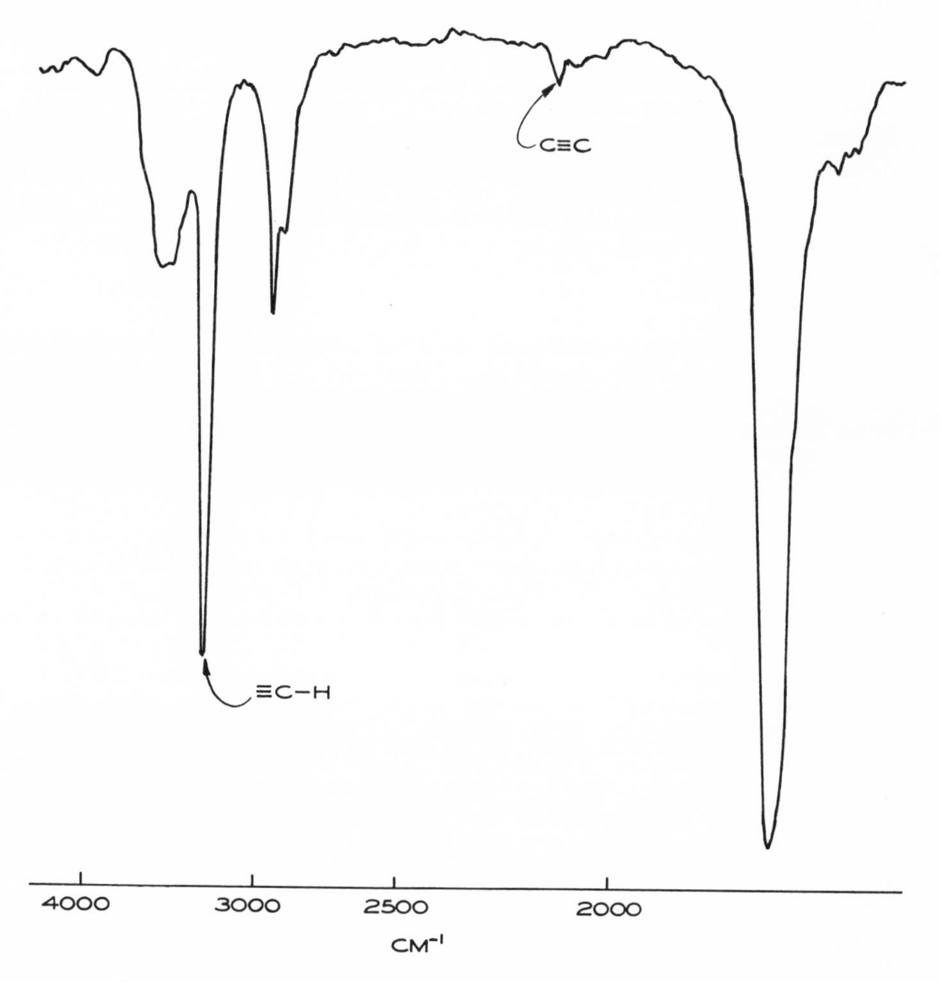

Figure 3.2. IR spectrum of 3-hexynyl acetate.

For Raman analysis, samples taken during the hydrogenation reaction to form an olefin are examined in the

1600-2300 cm^{-1} region. The percentage decrease in the C≡C band intensity is a rough measure of alkene formed; the (Z) double bond stretching band near 1650 cm^{-1} (see p.112) increases in intensity as the reaction proceeds.

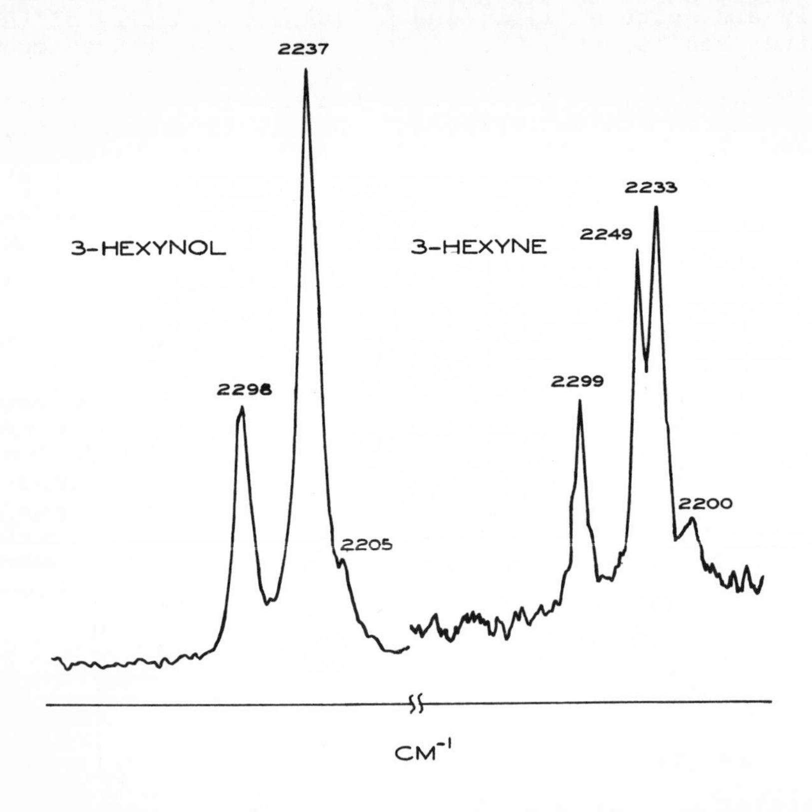

Figure 3.3. Raman spectra of 3-hexynol and 3-hexyne.

Alkyne, alkene, and alkane quantities may be estimated in the following way (6). An approximately equal mixture of olefin and acetylene is accurately prepared and its spectrum recorded between 1600 and 2300 cm^{-1} in order to obtain actual and relative intensities (peak height) of the two bands. The pertinent spectral region of the standard mixture is recorded immediately before examining the reaction products. Observation of an emission near 1670 cm^{-1} indicates that the (E), or *trans*, isomer is present. If alkane is produced along with alkene during the course of hydrogenation, the sum of the normalized C≡C and C=C band intensities will be less than expected. [Since the scattering efficiencies of (Z) and (E) double bonds in isomers usually are quite similar (see p.116), the quantity of the latter can be estimated.] In view of the errors accompanying band strength measurements (±3%), more than 5-10% alkane must form before the quantity of saturated hydrocarbon can be estimated. Unlike IR absorptions, Raman band intensities are dependent on the refractive index of the solvent; but since the difference between refractive indexes of alkyne and alkene is relatively small, no correction factor need be introduced for this type of analysis.

Reduction of dialkyl alkynes with alkali metals in liquid ammonia yields the (E) alkene exclusively; therefore, this reaction is a simpler one to monitor (5). Cycloalkynes behave differently, giving rise to mixtures of (Z) and (E) cycloalkenes in various proportions. In fact, reduction of I with sodium in liquid ammonia gives the (Z) olefin II as the sole product, proving that a mechanism of (Z) addition is not operative (7).

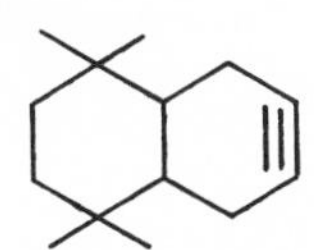

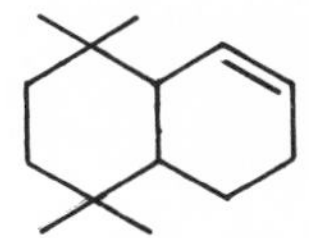

3.3 NITRILES

Nitriles generally are difficult to recognize by NMR and the intensity of the nitrile IR absorption varies from strong to undetectable. In unconjugated nitriles containing C, H, and N only, a medium intensity band appears between 2240-2260 cm^{-1} but the band becomes weak or absent when the α-carbon bears an electron attracting atom such as chlorine or oxygen (8). Consequently, the absence of an IR absorption near 2250 cm^{-1} cannot be taken as evidence for the absence of a C≡N group. In contrast, the strong Raman bands displayed by aliphatic nitriles persists when an electronegative substituent is situated on the α-carbon atom. A case in point is acetonitrile (Figure 3.4).

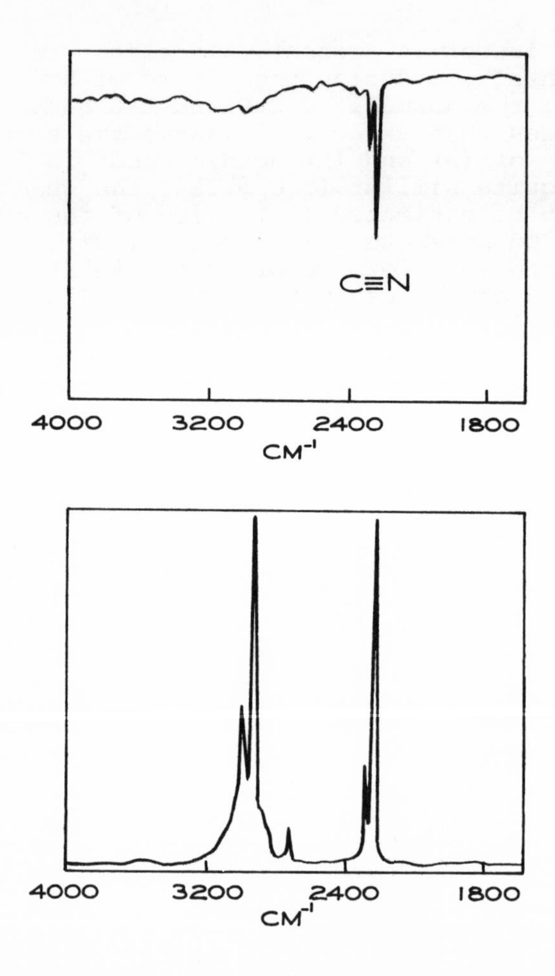

Figure 3.4. Infrared (top) and Raman (bottom) spectra of acetonitrile showing characteristic —C≡N stretch near 2250 cm^{-1}. (The band splitting is due to Fermi resonance.)

While its infrared C≡N stretching band strength is ten times that of chloroacetonitrile, Raman intensities for the two compounds are similar. In this context, Raman spectroscopy can be used advantageously to monitor the formation and dehydration of cyanohydrins, intermediates in the synthesis of α-unsaturated acids:

$$RCH_2-\overset{O}{\overset{\|}{C}}-R' \xrightarrow[NaCN]{H+} RCH_2\underset{CN}{\underset{|}{\overset{OH}{\overset{|}{C}}}}-R' \xrightarrow[-H_2O]{H+} RC=\overset{R'}{\overset{|}{C}}-COOH$$

Band frequencies and depolarization data for several nitriles are listed in Table 3.2 (6).

TABLE 3.2. Raman Spectral Data for Some Nitriles

Nitrile	ν (cm^{-1})	ρ
Aceto	2250	0.05
Proprio	2246	0.07
Isobutyro	2248	0.09
Citronellyl	2248	0.10
Geranyl	2226	0.14
Benzo	2228	0.17
Pyrazinecarbo	2228	0.19

Note that conjugation with ethylenic double bonds or aromatic rings increases ρ values compared with unconjugated nitriles. Conjugated nitriles display intense Raman bands near 2225 cm^{-1}, making it possible to determine small quantities in the presence of saturated analogs (Table 3.3) (6).

TABLE 3.3. Intensity Ratios of Some Conjugated/Unconjugated Nitriles

Equimolar Mixture	Intensity Ratio Conj./Unconj. C≡N
C=C-CN/C-C-CN	2.5
p-MeϕCN/ϕC-C-CN	8.0
ϕC=C-CN/ϕC-C-CN	25

REFERENCES

1. H. Sloane, Appl. Spectrosc. 25, 430 (1971).
2. E. Fermi, Z. Physik. 71, 250 (1931).
3. F. F. Cleveland and M. J. Murray, J. Am. Chem. Soc. 62, 3185 (1940).
4. F. F. Cleveland and M. J. Murray, J. Chem. Phys. 9, 390 (1941).
5. H. O. House, *Modern Synthetic Reactions* (Benjamin, New York, 1965).
6. S. K. Freeman (unpublished work).
7. M. Svoboda, J. Sicher, and J. Zavada, Tetrahedron Lett. 1, 15 (1964).
8. R. E. Kitson and N. E. Griffith, Anal. Chem. 24, 334 (1952).

Chapter 4

CARBONYLS

4.1 INTRODUCTION

It would be difficult to refute the statement that IR is the most sensitive spectral probe for carbonyls, but there are cogent reasons for using Raman spectroscopy. Apart from external factors, such as the occurrence of various forms of association, a host of internal ones may affect the frequency of the carbonyl group by perturbing its electronic distribution. Any factor tending to increase the force constant of the C=O band, that is, to increase the double bond character, will cause a frequency increase of the scattered or absorbed radiation. Conversely, any factor resulting in a diminution of the double bond character will lower the Raman or IR frequency. Considering the basic structure $\genfrac{}{}{0pt}{}{R}{X}{>}C{=}O$, the interaction between X and the carbonyl group is at a minimum when X is an alkyl group. The combined effects of induction (I) and mesomerism (M) for various X moieties are shown in Table 4.1. Other factors affecting the electron distribution, and therefore affecting the carbonyl stretching frequency, include steric, field, chelation, and conjugate chelation effects (Table 4.2). Steric hindrance to planarity (steric effect) takes place when bulky substituents render less effective the conjugation of a carbonyl group with an aromatic ring. Electrical interactions through space (field effect) can bring about an increased double bond character of the carbonyl moiety when electron attracting groups are carried by the α-carbon atom. The C=O stretching frequency of ortho-substituted aromatic compounds is lowered appreciably when the substituent is capable of forming a hydrogen bond with the carbonyl (chelation effect). Enolization of β-diketones and β-ketoacids accompanied by hydrogen bonding (conjugate chelation effect) causes a marked change in the electron distribution of the carbonyl group.

TABLE 4.1. Inductive (I) and Mesomeric (M) Effects on Carbonyl Stretching Frequencies

Carbonyl Type	Inductive and Mesomeric Effects	Net Effect on Double Bond Character	ν (cm^{-1}) (approximate)
Alkyl Ketone	R–C(=O)–R'	"Standard"	1715
α,β-Unsaturated Alkyl Ketone	R–C(=O)–C=C–R' (+M, -I)	Decreases	1670
Alkyl Acid Chloride	R–C(=O)–Cl	Increases	1800
Alkyl Ester	RC(=O)–O–R' (+M)	Increases	1740
Conjugated Alkyl Ester	RC=C–C(=O)–O–R'	Slightly Increases	1725
Alkyl Vinyl Ester	RC(=O)–O–C=C–R' (-I, +M)	Increases	1760
Alkyl Thiol Ester	RC(=O)–S–R'	Decreases	1690

The abnormally high ketone frequencies in strained ring compounds (Table 4.3) occur without a corresponding increase in the double bond character and are probably due to an altered interaction of the C=O bond with the adjacent C-C bond (1,2).

The foregoing comments make it abundantly clear that a carbonyl stretching vibrational frequency is an important interpretive clue. Carbonyl band intensities are considerably weaker in Raman than in IR spectra, but there is no difficulty in accurately determining Raman frequencies. In fact, modern Raman instrumentation usually permits the recording of a carbonyl band at full spectral scale. While the carbonyl frequency is a complex function of inductive, resonance, and other effects, Raman intensities appear to depend mainly on resonance effects, being little affected by inductive phenomena (3). An increase in band strength for conjugated carbonyls occurs in the order ester < ketone < aldehyde.

TABLE 4.2. Some Other Factors Affecting Carbonyl Frequencies

STERIC		FIELD	
H_3C–C(=O)–phenyl	H_3C–C(=O)–2,6-di-CH_3-phenyl	Cl, O, H, H, CH_3	H, O, Cl, H, CH_3
1690 cm^{-1}	1705 cm^{-1}	1743 cm^{-1}	1725 cm^{-1}
CHELATION		**CONJUGATE CHELATION**	
H_3C–C(=O)–3-OH-phenyl	H_3C–C(=O)–2-OH-phenyl (O···H–O)	R–C(=O)–CR_1R_2–C(=O)–R	R–C(OH)=CR_1–C(=O)–R_2
1685 cm^{-1}	1640 cm^{-1}	1715 cm^{-1}	1550–1650 cm^{-1}

4.2 DEPOLARIZATION VALUES

IR spectroscopy offers advantages over the Raman effect by virtue of characteristic absorptions in other regions of the spectrum, for example:

1. Ketones: A strong band occurs between 1350 and 1370 cm^{-1} for methyl ketones, ascribed to the symmetrical methyl deformation (1).
2. Aldehydes: The CH group in most aldehydes with alkyl groups or aromatic rings adjacent to the carbonyl gives rise to two Fermi resonance bands at 2800-2900 cm^{-1} and 2695-2775 cm^{-1} (1).
3. Esters: One or more strong absorption bands are observed between 1000 and 1250 cm^{-1} which can be utilized to identify the acid moiety (5).

TABLE 4.3. Effect of Ring Strain on Ketone C=O Stretching Frequencies

Ketone	$\nu_{C=O}$ (cm^{-1})
$(CH_3)_3C-C(=O)-C(CH_3)_3$	1686
$H_3C-C(=O)-CH_3$	1715
cyclohexanone	1717
cyclopentanone	1730-1750*
cyclobutanone	1782

*Fermi resonance doublet

4. Carboxylic Acids: Strongly bonded, very broad O-H stretching bands appear at about 3000 cm^{-1} (1).

Although such auxiliary band information generally is not available from Raman spectra, a unique characteristic of a Raman carbonyl band is the depolarization ratio (see p.26). Data listed in Tables 4.4-4.13 and summarized in Table 4.14 show the ρ value to be of considerable interpretive value (4). Some tentative conclusions can be drawn at the present time:

1. Conjugated dienic acyclic esters, ketones, and aldehydes are distinguishable.
2. It is possible to differentiate aromatic esters from aromatic ketones and aldehydes, endo-conjugated aldehydes from the analogous esters and ketones, and s-*trans* from s-*cis* acyclic and cyclic ketones.
3. In some instances the choice between a δ-lactone and an aliphatic ester can be properly accomplished. Interpreting an IR spectrum in this respect ordinarily is not a simple task.

4.2.1 Aldehydes

Acyclic and Cyclic Aldehydes (Table 4.4)

Analogous to ketones and esters, ρ values of acyclic saturated and α-unsaturated aldehydes are essentially the same. The degree of depolarization for these aldehydes falls within the relatively narrow range of 0.24-0.28. Endocyclic conjugated compounds exhibit somewhat higher ρ values (0.32) and an appreciable exaltation is observed for acyclic α,β-unsaturated aldehydes (about 0.4).

Aromatic Aldehydes (Table 4.5)

Although depolarization ratios for aldehydes conjugated with an aromatic ring (0.21-0.29) are similar to the acyclics (0.24-0.28), conjugation involving a disubstituted C=C and an aromatic nucleus such as cinnamic aldehyde gives rise to an appreciable ρ increase (0.32-0.40).

The carbonyl doublet of furfural (Figure 4.1) has been ascribed to an equilibrium mixture of two planar rotational isomers (6). Probably, proximity of the carbonyl group to the furan ring in I decreases the symmetry of the C=O stretching vibration, thereby increasing the degree of depolarization.

TABLE 4.4. Aldehyde Carbonyl Frequencies and Depolarization Values--Acyclic and Cyclic

Chemical Name*	Structure	$\nu_{C=O}$ (cm^{-1})	ρ
Propionaldehyde *VH2*	C–C–CHO	1726	0.28
Butyraldehyde *VH3*	C_2–C–CHO	1730	0.27
Hexanal *VH5*	C_4–C–CHO	1728	0.27
2,2-Dimethyl propionaldehyde *VHX*	C–C(C)(C)–CHO	1727	0.26
(Methylthio) acet-aldehyde *VH1S1*	C–S–C–CHO	1716	0.22
2-Phenyl propion-aldehyde *VHYR*	C–C(ϕ)–CHO	1727	0.26
2,7-Dimethyl-6-octen-1-al *VHY&4UY*	C–C(C)=C–C_3–C(C)–CHO	1727	0.28
2-Propenal *VH1U1*	C=C–CHO	1691	0.18
2-Methyl-2-propenal *VHYU1*	C=C(C)–CHO	1690	0.25
(E)-2-Butenal *VH1U2 -T*	C–C(H)=C(H)–CHO	1698	0.28
(E)-2-Hexenal *VH1U4 -T*	C_3C(H)=C(H)–CHO	1698	0.28
2-Methyl-2-butenal *VHYU2*	C–C=C(C)–CHO	1694	0.26
2-Methyl-2-pentenal *VHYU3*	C_2C=C(C)–CHO	1692	0.29

TABLE 4.4. (Continued)

Chemical Name*	Structure	$\nu_{C=O}$ (cm^{-1})	ρ
(E)-,(E)-2,4-Octadienal *VH1U2U4 -TT*	$C_3-\overset{H}{C}=\underset{H}{C}-\overset{H}{C}=\underset{H}{C}-CHO$	1667	0.4
(E)-,(E)-2,4-Decadienal *VH1U2U6 -TT*	$C_5-\overset{H}{C}=\underset{H}{C}-\overset{H}{C}=\underset{H}{C}-CHO$	1665	0.45
(E)-,(Z)-2,4-Decadienal *VH1U2U6 -TC*	$C_5-\overset{H}{C}=\overset{H}{C}-\overset{H}{C}=\underset{H}{C}-CHO$	1667	0.45
2,7-Dimethyl-2,6-octadienal *VHY&U4UY*	$C-\underset{C}{C}=C-C_2-C=\underset{C}{C}-CHO$	1682	0.24
3,7-Dimethyl-2-methylene-6-octenal *VHYU1&Y&3UY*	$C-\underset{C}{C}=C-C_2\overset{C}{C}-\underset{\overset{\parallel}{C}}{C}-CHO$	1690	0.32
4-Methyl-2-phenyl-2-pentenal *VHYR&U1Y*	$C-\underset{C}{C}-C=\underset{C_6H_5}{C}-CHO$	1690	0.33
4-Isopropyl-1-cyclohexene-1-carboxaldehyde *L6UTJ AVH DY*	CHO	1690	0.32
4-Isopropenyl-1-cyclohexene-1-carboxaldehyde *L6UTJ AVH DYU1*	CHO	1688	0.33
2-Pinen-10-al *L46 A EUTJ A A EVH*	CHO	1682	0.30

*Wiswesser Line Notation in italics

The same situation seems to exist for 2-thiophenecarboxaldehyde and its 5-methyl derivative. Expectedly, a single carbonyl band appears in the spectrum of 3-thiophenecarboxaldehyde.

TABLE 4.5. Aldehyde Carbonyl Frequencies and Depolarization Values--Aromatic

Chemical Name*	Structure	$\nu_{C=O}$ (cm^{-1})	ρ
Benzaldehyde *VHR*	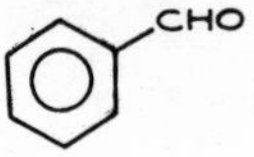	1705	0.24
Salicylaldehyde *VHR BQ*	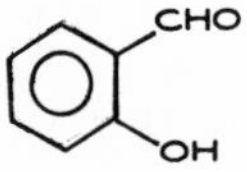	1675	0.28
p-Anisaldehyde *VHR DO1*	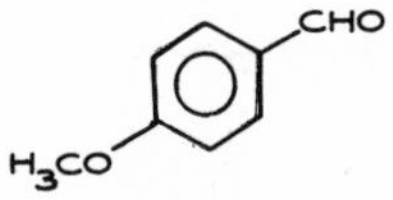	1695 1685	0.3 0.21
Cinnamaldehyde *VH1U1R -T*	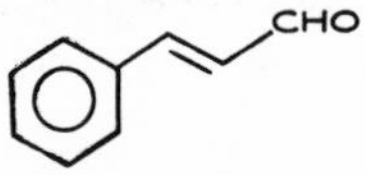	1668	0.32
α-Pentyl cinnamaldehyde *VHY5&U1R -T*	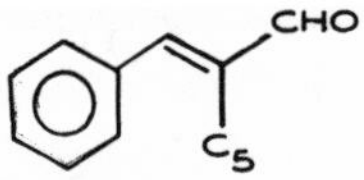	1667	0.38
α-Hexyl cinnamaldehyde *VHY6&U1R B -T*	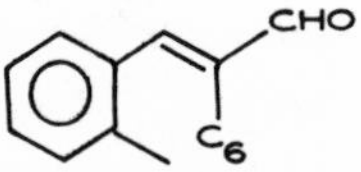	1668	0.40
2-Furaldehyde *T5OJ BVH*	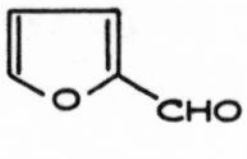	1692 1670	0.29 0.21
5-Methyl-2-furaldehyde *T5OJ BHV E*	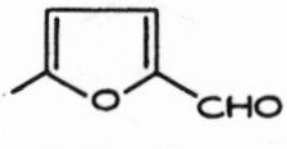	1671	0.24

The carbonyl region of pyrrole-2-carboxaldehyde shows one band only, but its half-band width (about 40 cm^{-1}) is approximately twice that of a normal carbonyl group, suggesting the presence of two unresolved bands.

TABLE 4.5 (Continued)

Chemical Name*	Structure	$\nu_{C=O}$ (cm^{-1})	ρ
2-Thiophenecarbox-aldehyde *T5SJ BVH*	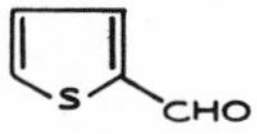	1675 1652	0.28 0.20
5-Methyl-2-thiophene-carboxaldehyde *T5SJ BVH E*	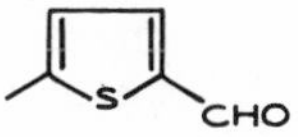	1672 1650	0.30 0.20
2,5-Dimethyl-3-thio-phene-carboxaldehyde *T5SJ B CVH E*	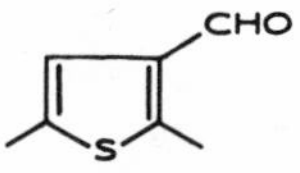	1680	0.21
Pyrrole-2-carbox-aldehyde *T5MJ BVH*	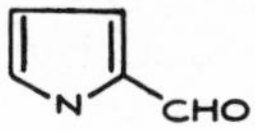	1645	0.28
3-Pyridinecarbox-aldehyde *T6NJ CVH*	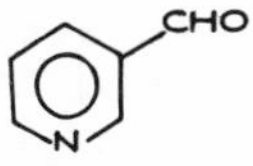	1702	0.25

*Wiswesser Line Notation in italics

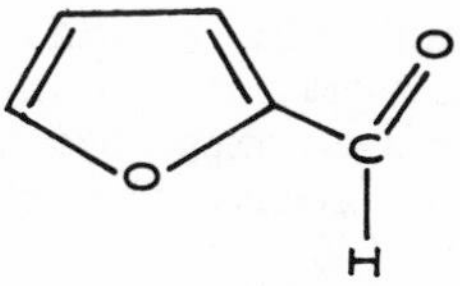

$\nu_{C=O} = 1692\ cm^{-1}$

$\rho = 0.30$

I

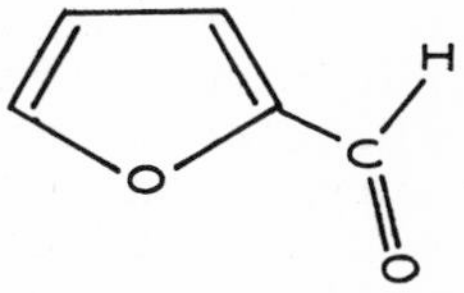

$\nu_{C=O} = 1672\ cm^{-1}$

$\rho = 0.19$

II

Figure 4.1. Rotational isomers of furfural.

4.2.2 Ketones

Acyclic and Cyclic Ketones (Table 4.6)

Acyclic ketones depolarize laser radiation to a lesser extent than the corresponding aldehydes (ρ = 0.22 compared to 0.27). Polarization measurements on cyclic ketones (ρ = 0.28) indicate that ρ values can be used to differentiate them from acyclic ketones. The low frequency component of the Fermi resonance doublet in cyclopentanone (7,8) and cyclohexanone has approximately the same ρ value as that of acyclic ketones ($\rho \sim 0.2$), but the high frequency band possesses a larger depolarization ratio ($\rho \sim 0.3$). This behavior is quite similar to the Fermi resonance bands observed in conjugated endocyclic ketones. Attention is drawn to the unusually high ρ values of the three listed cyclobutanones (ρ = 0.4-0.5).

Vibrational spectral studies on α-haloketones (9) indicate that these molecules exist in two conformational states, as evidenced by a splitting of carbonyl and carbon-halogen stretching bands. In the case of 1-chloro-2-propanone (Figure 4.2), the chlorine atom oriented *trans* to the methyl group will inhibit C=O resonance, increasing the double bond character, thereby raising the carbonyl stretching frequency. The *trans* conformer is preferred on steric grounds because the repulsion between a chlorine and an oxygen is considerably less than that between a chlorine atom and a methyl group. Electrostatic interaction between the chlorine and oxygen atoms tends to force the molecule from a *cis* to a *gauche* arrangement. It is believed that the higher frequency IR band has its basis in a field effect operating through space, rather than in an inductive effect transmitted through the bonds (9). The chlorine atom in the *trans* rotamer may decrease the symmetry of the C=O stretching vibration compared to the unperturbed carbonyl in the *gauche* form, thereby bringing about an increase in ρ value (4).

α-Unsaturated Ketones (Table 4.7)

The existance of s-*cis*, *trans* isomerism in acyclic α-unsaturated ketones (Figure 4.3), where "s" signifies limited rotation about the single bond joining the C=O and C=C moieties, arises from steric interactions involving R, R_1, R_2, and R_3.

TABLE 4.6. Ketone Carbonyl Frequencies and Depolarization Values--Saturated Acyclic and Cyclic

Chemical Name*	Structure	$\nu_{C=O}$ (cm^{-1})	ρ
Acetone *1V1*	$C-\overset{O}{\overset{\|}{C}}-C$	1715	0.20
2-Butanone *2V1*	$C-\overset{O}{\overset{\|}{C}}-C_2$	1714	0.20
3-Hexanone *3V2*	$C_2-\overset{O}{\overset{\|}{C}}-C_3$	1716	0.21
Chloro-2-propanone *G1V1*	$Cl-C-\overset{O}{\overset{\|}{C}}-C$	1743 1725	0.33 0.25
Cyclobutanone *L4VTJ*		1785	0.42
4,7,7-trimethyl-bicyclo[3.2.0]hept-3-en-6-one *L45 BV FUTJ C C G*	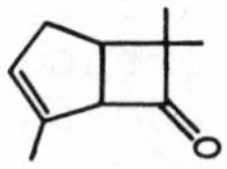	1775	0.39
3,3-dimethylcyclo-butanone *L4VTJ C C*	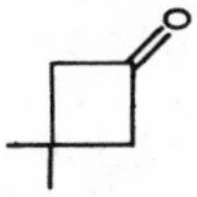	1782	0.50
Cyclopentanone *L5VTJ*	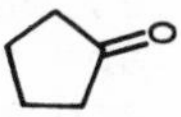	1745 1730	0.30 0.20
Cyclohexanone *L6VTJ*	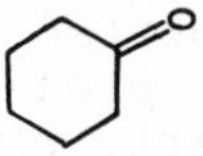	1728 1715	0.30 0.20
Cycloheptanone *L7VTJ*	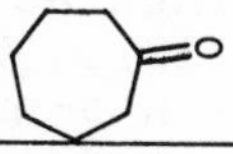	1700	0.27

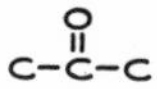

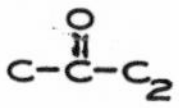

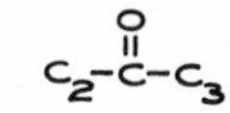

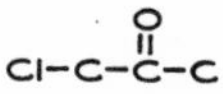

*Wiswesser Line Notation in italics

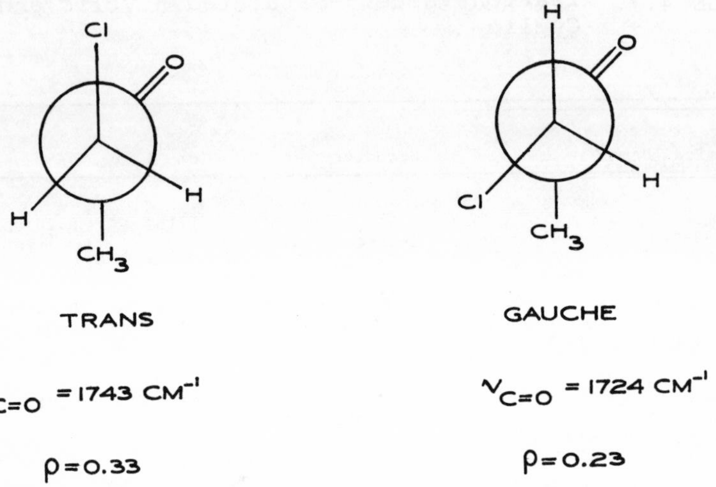

TRANS

$\nu_{C=O} = 1743\ CM^{-1}$

$\rho = 0.33$

GAUCHE

$\nu_{C=O} = 1724\ CM^{-1}$

$\rho = 0.23$

Figure 4.2. Rotational isomers of 1-chloro-2-propanone.

For those cases where R=H and R, R_2 and R_3=alkyl, there will be less steric interference between R and H in the s-*cis* form than between R and R_3 in the s-*trans* conformation, leading to a preference for the s-*cis* rotamer (10). (Cyclic molecules, on the other hand, are fixed cisoid and transoid.) The presence of other preferred, nonplanar conformations adopted by some acyclic conjugated ketones has been inferred from IR (11) and NMR (12) data. For example, when R and R_2 are as large as a combination of ethyl and methyl, deviation from planarity is significant, and when R_3 also is an alkyl group it appears likely that the preferred conformation has the carbonyl group rotated about 90° out of the carbon-carbon double bond plane. The decrease in energy due to a lessening in steric crowding more than compensates for the energy decrease attributed to a lower degree of orbital overlap interaction of the two groups in a planar arrangement (12). The s-*cis* and s-*trans* conformations of open chain α-unsaturated ketones can be differentiated by the locations of C=O and C=C Raman and IR stretching bands (10,11,13,15).

TABLE 4.7. Ketone Carbonyl Frequencies and Depolarization Values--α-Unsaturated

Chemical Name*	Structure	$\nu_{C=O}$ (cm^{-1})	ρ
(E)-3-Penten-2-one *2U1V1*	C-C(H)=C(H)-C(=O)-C	1673	0.21
(E)-4-Hexen-3-one *2V1U2*	C-C(H)=C(H)-C(=O)-C_2	1674	0.22
3-Methyl-3-penten-2-one *2YV1*	C-C=C(C)-C(=O)-C	1669	0.21
4-Methyl-3-penten-2-one *1YU1V1*	C-C(C)=C-C(=O)-C	1670	0.40
3-Buten-2-one *1V1U1*	C=C-C(=O)-C	1670	0.14
3-Methyl-3-buten-2-one *1VYU1*	C=C(C)-C(=O)-C	1672	0.21
4-Hexyn-3-one *2V1UU2*	C-C≡C-C(=O)-C_2	1675	0.18
6-Methyl-(E)-3,5-heptadien-2-one *1YU2U1V1*	C-C(C)=C-C(H)=C(H)-C(=O)-C	1630	0.24
1-(2,6,6-Trimethyl-2-cyclohexen-1-yl)-1-penten-3-one *L6UTJ A E E F1U1V2 -T*		1670	0.18[a]
4-(2,6,6-Trimethyl-2-cyclohexen-1-yl)-3-buten-2-one *L6UTJ A E E FUV1 -T*		1670	0.16

TABLE 4.7. (Continued)

Chemical Name*	Structure	$\nu_{C=O}$ (cm^{-1})	ρ
3-Methyl-2-cyclohexen-1-one *L6V BUTJ C*	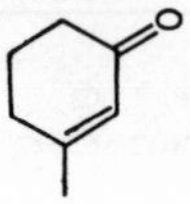	1677	0.28 0.21
3,5-Dimethyl-2-cyclohexen-1-one *L6V BUTJ C E*	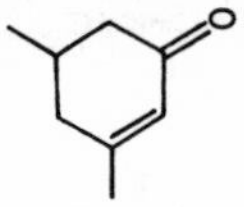	1672	0.29 0.22
2,4,4-Trimethyl-2-cyclohexen-1-one *L6V BUTJ B D D*	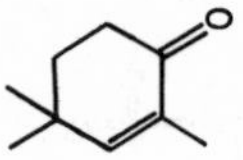	1670	0.23
p-Mentha-6,8-dien-2-one *L6V BUTJ B EYU1*	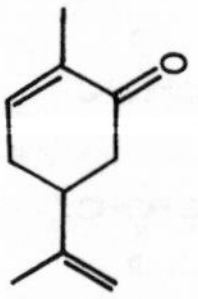	1622	0.23
p-Menth-1-en-3-one *L6V BUTJ C FY*	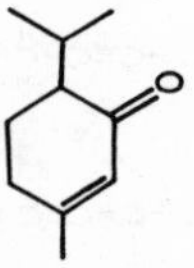	1673	0.23
2-Pinen-4-one *L46 A EV FUTJ A A G*	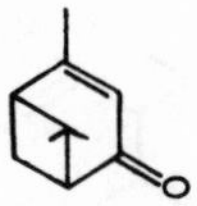	1680	0.21

The s-*cis* conformation is characterized by a severe diminution in C=O band strength compared to the corresponding s-*trans* form. The frequency difference between C=O and C=C stretching modes are characteristic of the conformation (16): s-*cis* compounds > 70 cm^{-1} and s-*trans* < 60 cm^{-1}.

TABLE 4.7. (Continued)

Chemical Name*	Structure	$\nu_{C=O}$ (cm^{-1})	ρ
3-Methyl-4-(2,6,6-Tri-methyl-2-cyclohexen-1-yl)-3-buten-2-one *L6UTJ A E E F1UYV1 -T*	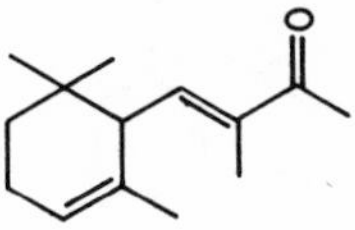	1665	0.22
4-(2,6,6-Trimethyl-1-cyclohexen-1-yl)-3-buten-2-one *L6UTJ A B1U1V1 C C -T*	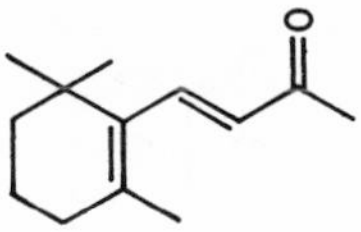	1667	0.27
4-(2-Furyl)-3-buten-2-one *T5OJ B1U1V1 -T*	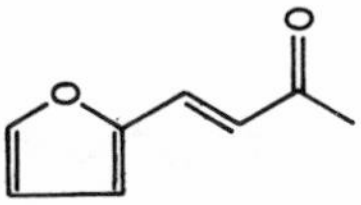	1672	0.26
3-Methyl-4-(2-furyl)-3-buten-2-one *T5OJ B1UYV1 -T*	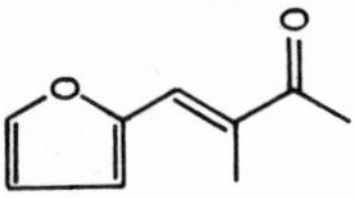	1670	0.35
2-Cyclopenten-1-one *L5V BUTJ*	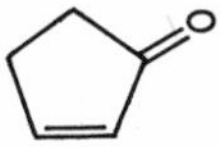	1713	0.26
2-Cyclohexen-1-one *L6V BUTJ*	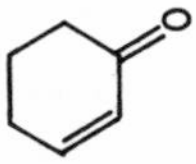	1685	0.30 0.22
6-Methyl-2-cyclohexen-1-one *L6V BUTJ F*	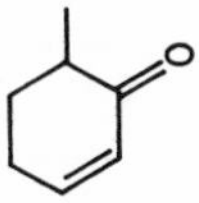	1680	0.22

This correlation does not hold for five-membered rings because strain effects result in a lowered C=C frequency and a raised C=O frequency.

The compounds shown in Figure 4.4 (4) adopt either s-*cis* or essentially s-*trans* forms.

TABLE 4.7. (Continued)

Chemical Name*	Structure	$\nu_{C=O}$ (cm^{-1})	ρ
6,7-Dihydro-1,1,2,3,3-pentamethyl-4(5H)-indanone *L56 FV AU- ETJ B B C D D*	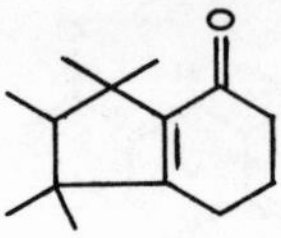	1669	0.24
Cedr-8-en-10-one *L B556 A KV IUTJ C G G I*	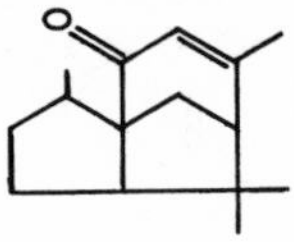	1673	0.22
5-(3,3-dimethyl-2-norbornylidene)-3-penten-2-one *L55 A CYTJ CU2U1V1 D D -T*	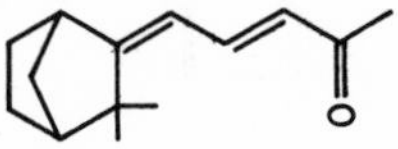	1637	0.23
2-Hydroxy-2-cyclohexen-1-one *L6V BUTJ BQ*	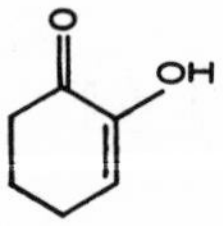	--[b]	--
2-Hydroxy-p-mentha-1-en-3-one *L6V BUTJ BQ C FY*	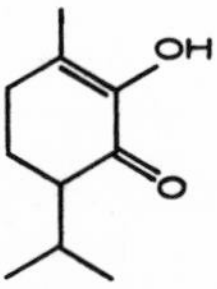	1668	0.20
Heptylidene cyclopentanone *L5VYTJ BU7*	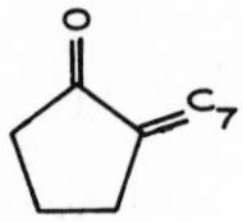	1715	0.56

The evidence presented indicates that depolarization ratios can be used to characterize these conformational states. Data for essentially s-*trans* acyclic molecules are similar to those observed for saturated acyclics (ρ ∿ 0.2), but depolarization values of molecules constrained in the s-*cis* conformation are considerably higher (ρ ∿ 0.4) and approach those of s-*cis* cyclic compounds (ρ = 0.42-0.52).

TABLE 4.7. (Continued)

Chemical Name*	Structure	$\nu_{C=O}$ (cm^{-1})	ρ
2-Heptylidene cyclohexanone *L6VYTJ BU7*	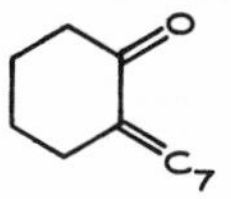	1691	0.55
p-Menth-4(8)-en-3-one *L6VYTJ BUY E*	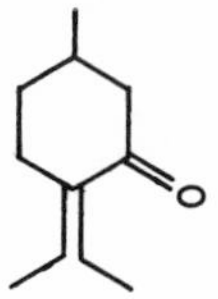	1682	0.51

*Wiswesser Line Notation in italics
[a]Combined C=C-H and C=O
[b]Very weak C=O band

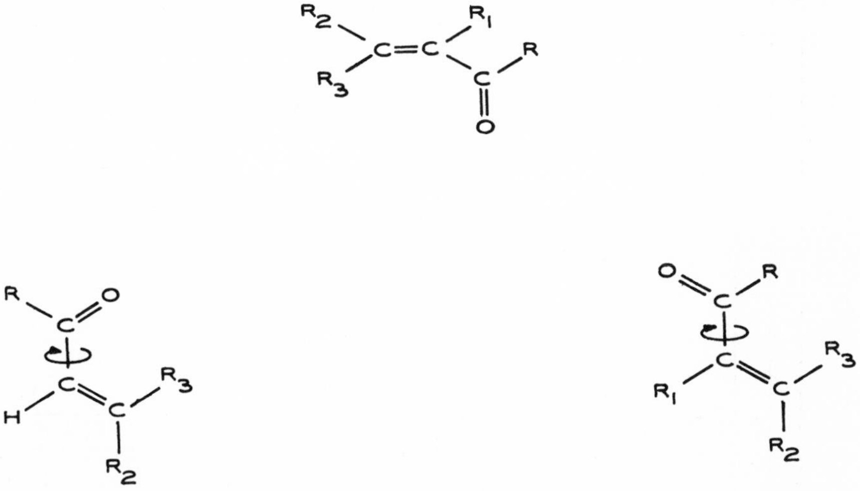

Figure 4.3. s-*cis* and s-*trans* Rotational isomers of α-unsaturated acyclic ketones. s-*cis* preferred to s-*trans* when R=H and R_1, R_2, R_3=alkyl.

Acyclic				Cyclic			
s-cis	ρ	s-trans	ρ	s-cis	ρ	s-trans	ρ
[structure]	0.40	[structure]	0.22	[structure]	0.46	[structure]	0.22
[structure]	0.42	[structure]	0.22	[structure]	0.52	[structure]	0.23

Figure 4.4. Depolarization values for s-*cis* and s-*trans* forms of some α-unsaturated ketones.

The considerable differences between cisoid and transoid conformers might be explained on the basis of vibrational coupling. The conjugated C=O and C=C groups are strongly coupled in s-*cis* ketones whereas they are more isolated in s-*trans* compounds. Consequently, the polarizability ellipsoid in the s-*cis* form is spread over four atoms compared to two in the case of s-*trans* molecules. The coupled C=O vibration would be expected to be less symmetric (higher ρ value) than the isolated stretching mode (lower ρ value). An alternate and more appealing explanation for the relatively large differences in degree of depolarization, which is in agreement with the preceding comments on aromatic aldehydes and 1-chloro-2-propanone, invokes a field effect (4). The proximity of the C=C to the C=O in the s-*cis* conformation could contribute to an increase in the asymmetry of the C=O stretching vibration compared to the *trans* arrangement. From a consideration of either of the two hypotheses, the s-*trans* conformers would be expected to display ρ values similar to the saturated molecules, in agreement with the experimental evidence. The slightly lower figures for the acyclic cisoids relative to the corresponding cyclics might be a reflection of a small degree of aplanarity in the acyclic molecules.

Those molecules existing as a mixture of s-*cis* and s-*trans* conformers display rather complex spectra between about 1600 and 1700 cm^{-1}. The Raman and IR curves for 3-octen-2-one in this region are shown in Figure 4.5 (4) and the interpretation is in agreement with that for 3-penten-2-one (10). These combined patterns, in addition to the depolarization values for the C=O bands, are diagnostic for conformeric mixtures of this general type. An important spectral feature is the extremely weak Raman line (1700 cm^{-1}, $\rho \sim 0.4$) for the s-*cis* rotamer. The C=C line for the s-*cis* rotational isomer is the low frequency component (1629 cm^{-1}) of the doublet. The intense C=O band, ascribed to the more abundant s-*trans* conformer (1677 cm^{-1}, $\rho = 0.22$) is accompanied by the C=C stretch observed for this form at 1642 cm^{-1}. The IR frequencies are in good agreement.

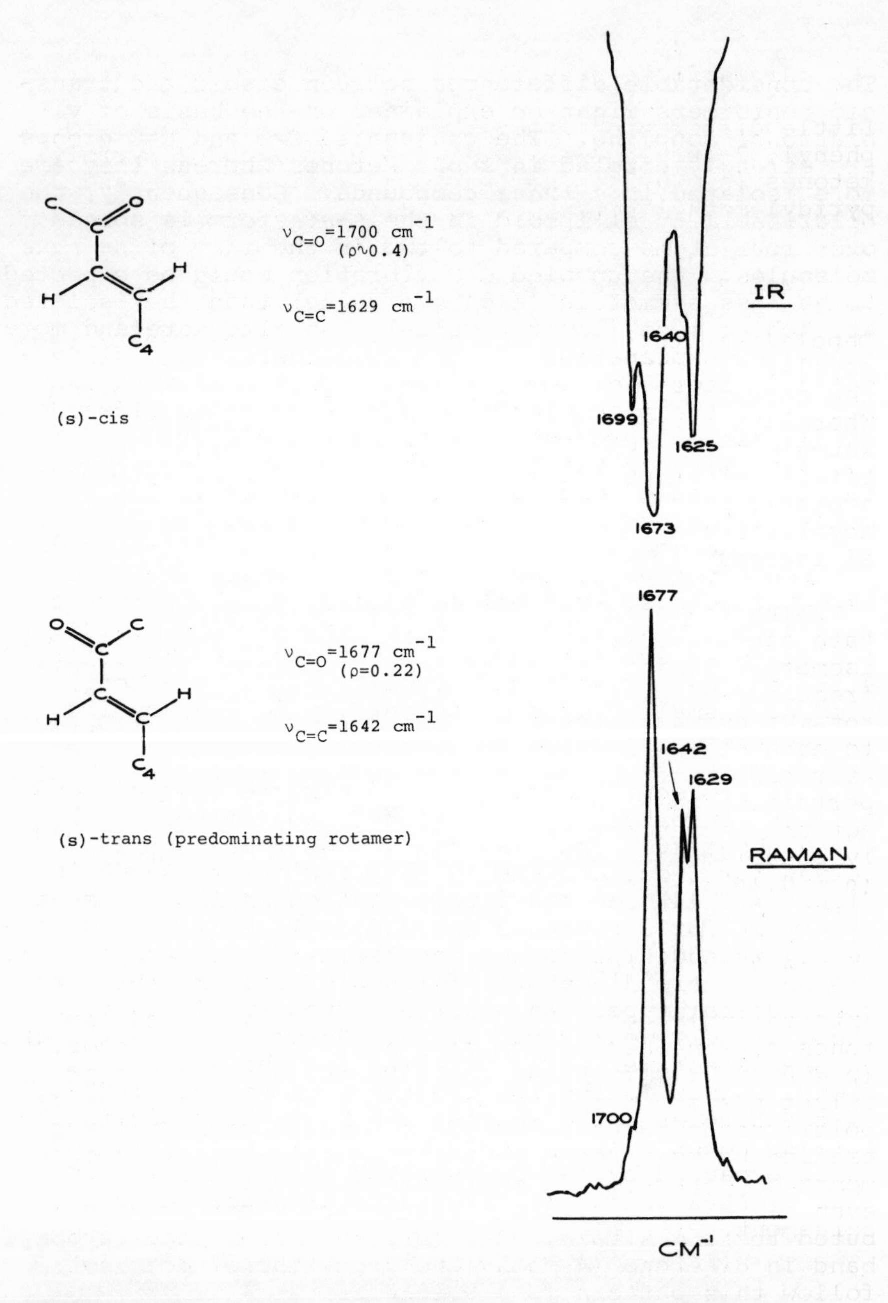

Figure 4.5. Raman and IR spectra of 3-octen-2-one.

Aromatic Ketones (Table 4.8)

Little difference is observed among the ρ values for phenyl, 2-furyl, 2-thienyl, 3-pyridyl, and 2-pyrrolyl ketones (ρ = 0.2). However, the ratios for 2- and 3-pyridyl methyl ketone are significantly lower (ρ ∿ 0.17).

4.2.3 Esters (Table 4.9)

Depolarization values for saturated and α-unsaturated acyclic esters (ρ ∿ 0.1) are considerably lower than the corresponding aldehydes and ketones (ρ = 0.2-0.3). Whereas a vinyl group conjuncted to an aldehydic or ketonic moiety severely decreases the ρ value, such an effect is not observed in esters. Conjugation with an aromatic ring raises the degree of depolarization over acyclic analogues and diconjugation causes an additional increase (ρ = 0.2).

Raman (17) and IR (18) spectra of ethyl chloroacetate each display two C=O bands arising from rotational isomers. Similar to 1-chloro-2-propanone, the higher frequency band (1762 cm^{-1}) has been assigned to the rotamer where the carbonyl and halogen are in proximity to each other, and the lower frequency band (1740 cm^{-1}) ascribed to the *gauche* form. In view of the discussion pertaining to 1-chloro-2-propanone (see p.86), it is not unexpected that the latter form scatters more highly polarized radiation (ρ = 0.10) than the former (ρ = 0.14).

4.2.4 Lactones (Table 4.10)

Interestingly, five- and six-membered saturated lactones have higher values (ρ = 0.15-0.17) than esters (ρ = 0.09-0.12). Since the carbonyl frequencies of δ-lactones and normal esters are very similar, the depolarization value is a good diagnostic for differentiating between these groups. Depolarization measurements for saturated C_4 and unsaturated C_5 rings are even higher (ρ = 0.22), a fact which might be attributed to ring strain. However, the bifurcated carbonyl band in diketene (4-methylene-2-oxetanone) does not follow this pattern (ρ = 0.15, 0.18).

TABLE 4.8. Ketone Carbonyl Frequencies and Depolarization Values--Aromatic

Chemical Name*	Structure	$\nu_{C=O}$ (cm^{-1})	ρ
Acetophenone *1VR*		1690	0.21
4-Methyl acetophenone *1VR D*		1681	0.20
2-Furyl methyl ketone *T5OJ BV1*		1685	0.20
Methyl 5-methyl-2-furyl ketone *T5OJ BV1 E*		1687	0.23
Methyl 2-thienyl ketone *T5SJ BV1*		1670	0.21
1-(2-Thienyl)-1-pentanone *T5SJ BV4*		1665	0.21
Methyl 5-methyl-2-thienyl ketone *T5SJ BV1 E*		1668	0.23

TABLE 4.8. (Continued)

Chemical Name*	Structure	$\nu_{C=O}$ (cm^{-1})	ρ
Methyl 3-pyridyl ketone *T6NJ CV1*	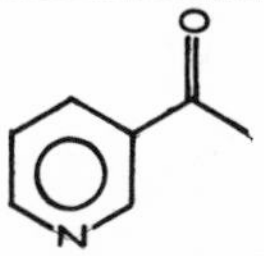	1695	0.19
Methyl 2-pyridyl ketone *T6NJ BV1*	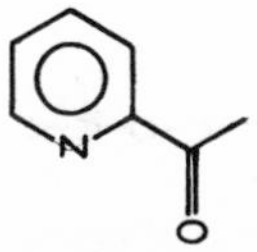	1710	0.17
Methyl 4-pyridyl ketone *T6NJ DV1*	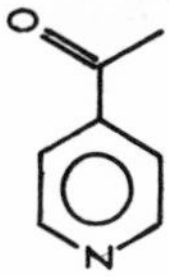	1698	0.16
Methyl pyrrol-2-yl ketone *T5NJ BV1*	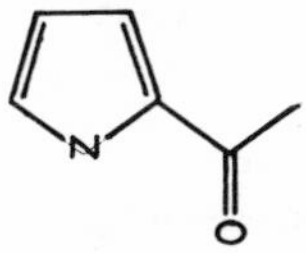	1640 1652 sh	0.24

*Wiswesser Line Notation in italics

4.2.5 Carboxylic Acids (Table 4.11)

The C=O stretching vibrations of these highly polar compounds evidence the lowest ρ value of all types studied (ρ = 0.05).

TABLE 4.9. Ester Carbonyl Frequencies and Depolarization Values

Chemical Name*	Structure	$\nu_{C=O}$ (cm^{-1})	ρ
Ethyl formate *VHO2*	HC(=O)-O-C_2	1725	0.10
Ethyl acetate *2OV1*	C-C(=O)-O-C_2	1740	0.11
Ethyl propionate *2OV2*	C_2C(=O)-O-C_2	1738	0.10
(E)-2-Hexen-1-yl acetate *2U2OV1*	C-C(H)=C(H)-C-O-C(=O)-C	1740	0.08
Ethyl acrylate *2OV1U1*	C=C-C(=O)-O-C_2	1726	0.08
Vinyl acetate *1VO1U1*	C=C-O-C(=O)-C	1760	0.09
Methyl mercapto-acetate *SH1VO1*	HS-C-C(=O)-O-C	1735	0.09
Methyl chloro-acetate *G1VO1*	Cl-C-C(=O)-O-C	1762 1740	0.14 0.10
Diethyl carbonate *2OVO2*	C_2-O-C(=O)-O-C_2	1745	0.05
Ethyl 3,7-dimethyl-2,6-octadienoate *2OV1UY&3UY*	C-C(C)=C-C-C-C(C)=C-C(=O)-O-C_2	1734	0.10
Diethyl maleate *2OV1U1VO2 -C*	C(-C(=O)-OC_2)=C(-C(=O)-OC_2)	1730	0.11

TABLE 4.9. (Continued)

Chemical Name*	Structure	$\nu_{C=O}$ (cm^{-1})	ρ
Diethyl fumarate *2OV1U1VO2 -T*	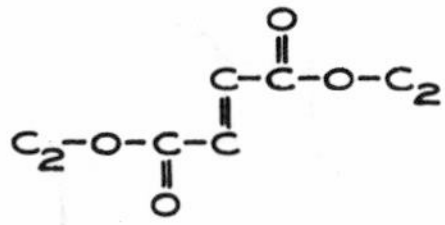	1730	0.12
Methyl cinnamate *1OV1U1R -T*	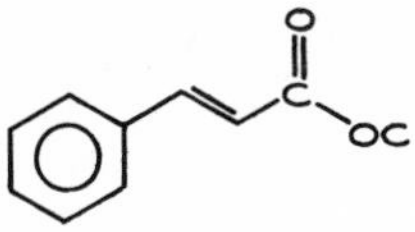	1709	0.20
Ethyl cinnamate *2OV1U1R -T*	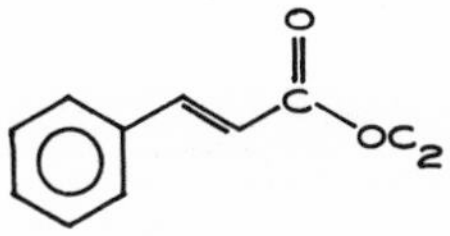	1706	0.21
Ethyl benzoate *2OVR*	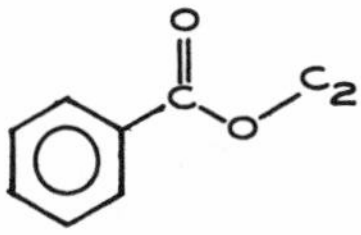	1720	0.15
Ethyl p-hydroxy benzoate *QR DVO2*	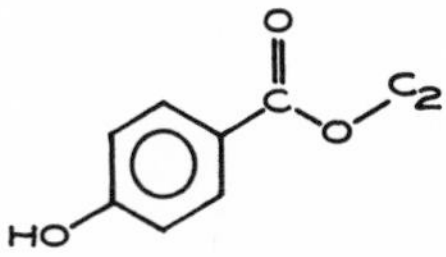	1715	0.18

*Wiswesser Line Notation in italics

In the case of conjugated acids, the C=C band swamps the C=O band, the latter usually appearing as a barely discernible shoulder. The data were gained on dimers, that is, spectra were recorded on neat liquids. Only the symmetric C=O stretching vibration of the centrosymmetric dimer appears in the Raman effect (Figure 4.6).

TABLE 4.10. Lactone Carbonyl Frequencies and Depolarization Values

Chemical Name*	Structure	$\nu_{C=O}$ (cm^{-1})	ρ
4-Methyl-2-oxetanone *T4OVTJ D*	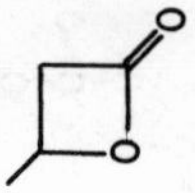	1828	0.23
Dihydro-2(3H)-furanone *T5OVTJ*	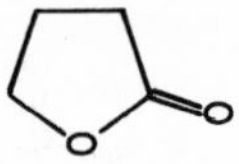	1780	0.15
5-Ethyl-dihydro-2(3H)-furanone *T5OVTJ E2*	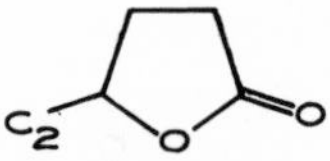 C_2	1782	0.14
5-Butyl-dihydro-2(3H)-furanone *T5OVTJ E4*	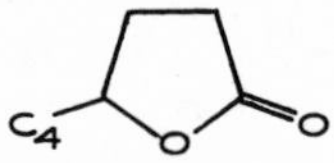 C_4	1779	0.17
Tetrahydro-2H-pyran-2-one *T6OVTJ*	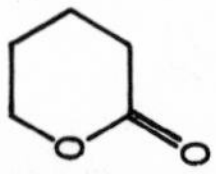	1745	0.15
Tetrahydro-6-methyl-2H-pyran-2-one *T6OVTJ F*	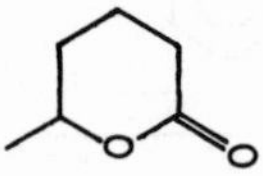	1742	0.15
4-Methylene-2-oxetanone *T4VOY DHJ CU1*	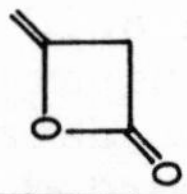	1895 1858	0.15 0.18

*Wiswesser Line Notation in italics

TABLE 4.11. Carboxylic Acid Carbonyl Frequencies and Depolarization Values

Chemical Name*	Structure	$\nu_{C=O}$ (cm^{-1})	ρ
Acetic acid *QV1*	C-C(=O)-OH	1670	0.04
Propionic acid *QV2*	C_2-C(=O)-OH	1672	0.04
Butyric acid *QV3*	C_3-C(=O)-OH	1670	0.05
(E)-2-Pentenoic acid *QV1U3 -T*	C_2C=C-C(=O)-OH	1632	<0.05
(E)-2-Hexenoic acid *QV1U4 -T*	C_3C=C-C(=O)-OH	1630	<0.05

*Wiswesser Line Notation in italics

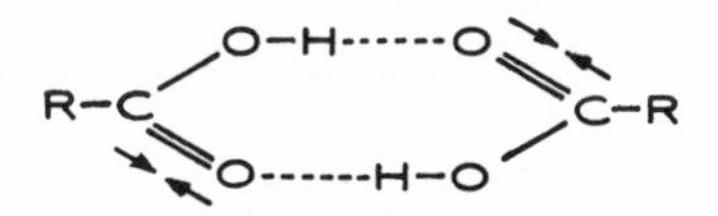

Figure 4.6. Symmetric carbonyl stretching vibration of carboxylic acids. (In the IR, the antisymmetric C=O stretch appears as an intense band near 1700 cm^{-1}.) $\nu_{C=O}$ ∿1670 cm^{-1} (very weak) ρ ∿0.05.

Carboxylic acids, excepting conjugated aromatics, are easily distinguished from other carbonylics by virtue of their extremely weak bands and very low depolarization values.

4.2.6 Carboxylic Acid Halides (Table 4.12)

An increase in electronegativity of the halogen atom is accompanied by an appreciable decrease in depolarization ratio.

TABLE 4.12. Carboxylic Acid Halide Carbonyl Frequencies and Depolarization Values

Chemical Name*	Structure	$\nu_{C=O}$ (cm^{-1})	ρ
Acetyl fluoride *FV1*	C-C(=O)-F	1845	∿0.1
Acetyl chloride *GV1*	C-C(=O)-Cl	1785	0.34
Acetyl bromide *EV1*	C-C(=O)-Br	1795	0.65

*Wiswesser Line Notation in italics

4.2.7 Ketoesters (Table 4.13)

A depolarization value of 0.16 for the single carbonyl band of ethyl pyruvate corresponds to the expected average of ketone and ester moieties.

TABLE 4.13. Ketoester Carbonyl Frequencies and Depolarization Values

Chemical Name*	Structure	$\nu_{C=O}$ (cm^{-1})	ρ
Pyruvic acid: ethyl ester *2OVV1*	C-C(=O)-C-O-C_2	1734	0.16
Acetoacetic acid: methyl ester *1V1VO1*	C-C(=O)-C-C(=O)-O-C	1750 1725	0.11 0.31
Acetoacetic acid: ethyl ester *2OV1V1*	C-C(=O)-C-C(=O)-O-C_2	1751 1725	0.12 0.35
Acetoacetic acid, 2-methyl: ethyl ester *2OVYV1*	C-C(=O)-C(C)-C(=O)-O-C_2	1750 1723	0.15 ∿0.40

*Wiswesser Line Notation in italics

A different situation exists for α-ketoesters. The higher values gained on the resolved bands indicate that, whereas the unchelated ketonic group in acetoacetic acid esters ($\rho \sim 0.3$) is affected by the presence of the ester carbonyl ($\rho = 0.11$), no reciprocal effect is observed. On the other hand, the 2-methyl derivative shows ρ enhancement for both carbonylics ($\rho = 0.4$ and 0.15).

4.3 SUMMARY

A summary of the depolarization data for the carbonyls discussed in 4.2.1-4.2.7 appears in Table 4.14.

4.4 INTENSITY RATIOS OF α-UNSATURATED CARBONYLS

Unlike IR, the Raman intensity ratio of C=O and C=C stretching bands in a conjugated carbonyl sometimes provides insights into the carbonyl type. Interestingly, relative intensity values obtained by different methods are in fair agreement (Table 4.15).

TABLE 4.14. Depolarization Values of Carbonyls--Summary of Tables 4.4-4.13

Compound Type	Acyclic	Alicyclic	Acyclic Conj. Unsubst.	Acyclic Conj. Subst.	Acyclic Conj. (dienic)	Cyclic Conj. Endo	Cyclic Conj. Exo	Aromatic
Aldehyde	0.27	--	0.27[a]	0.26[b]	0.38-0.45[c]	0.31	--	0.25±0.04
Ketone	0.21	0.28[d]	0.20[e]	0.21[f]	0.25	0.23±0.03	∿0.45	0.20±0.04
Ester	0.10±0.02	--	0.09	0.10	0.20	0.20	--	0.17±0.03
Carboxylic Acid	∿0.05	--	<0.05	--	--	--	--	--
Lactone	--	0.14-0.26	--	--	--	--	--	--
Acid Chloride	0.1-0.65	--	--	--	--	--	--	--
Ketoester α	0.16[g]							
β C=O	0.30	--	--	--	--	--	--	--
β C-O-C (C=O)	0.10							

[a] Vinyl--0.18.

[b] Exomethylene--0.32.

[c] Ranges given when ρ values fall outside experimental error (±10%).

[d] Cyclobutanones--0.4 to 0.5.

[e] Vinyl--0.14.

[f] Value for s-trans s-cis ∿0.4.

TABLE 4.15. Relative Intensities of C=C/C=O Raman Bands in Conjugated Carbonyls

Compound	$k_{C=C}/k_{C=O}$ [a]	$I_{C=C}/I_{C=O}$ [b]
Esters:		
C=C–C(=O)–O–C_2	2.0	2.8
C=C–O–C(=O)–C	-	5.0
C–CH=CH–C(=O)–O–C_2	1.8	2.0
C=C(C)–C(=O)–O–C_2	1.4	-
C_6H_5–C=C–C(=O)–OC_2	-	7.0
(cyclohexenyl)–C–C(=O)–OC_2	-	4.0
Ketones:		
C=C–C(=O)–C	0.85	1.0

TABLE 4.15. (Continued)

Compound	$k_{C=C}/k_{C=O}$ [a]	$I_{C=C}/I_{C=O}$ [b]
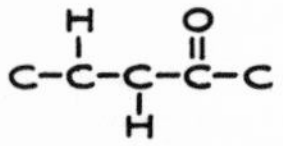 C–C(H)–C(H)–C(=O)–C	0.70	1.1
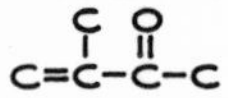 C=C(C)–C(=O)–C	0.75	1.0
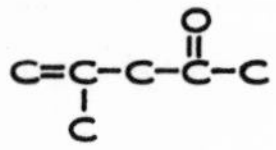 C=C(C)–C–C(=O)–C	7.0	10
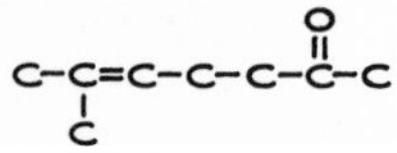 C–C(C)=C–C–C–C(=O)–C	-	5.0
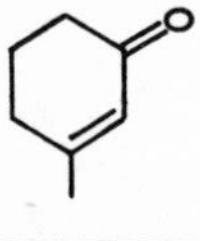	-	1.3
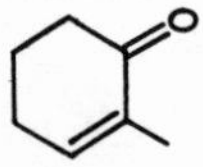	-	1.0
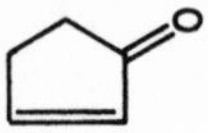	-	1.7

TABLE 4.15. (Continued)

Compound	$k_{C=C}/k_{C=O}$ [a]	$I_{C=C}/I_{C=O}$ [b]
Aldehydes: $C{=}C{-}CHO$	0.5	0.6
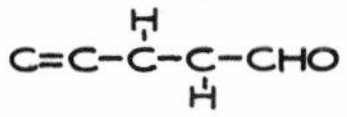 $C{=}C{-}\overset{H}{C}{-}\underset{H}{C}{-}CHO$	0.7	0.9
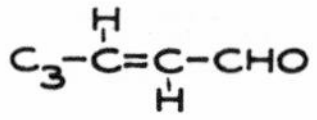 $C_3{-}\overset{H}{C}{=}\underset{H}{C}{-}CHO$	-	0.8
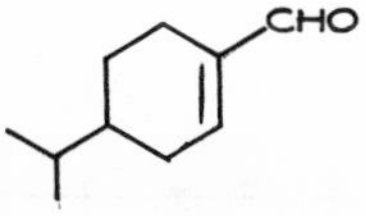 CHO	-	0.8
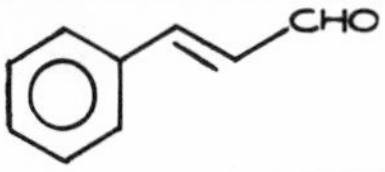 CHO	-	6.0
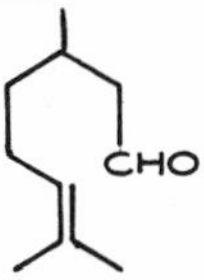 CHO	-	5.0
Acids: 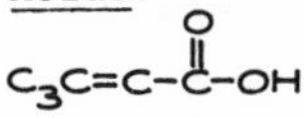 $C_3C{=}C{-}\overset{O}{\overset{\Vert}{C}}{-}OH$	-	20
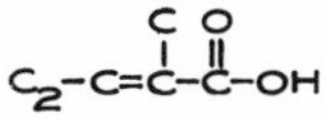 $C_2{-}C{=}\overset{C}{\overset{\vert}{C}}{-}\overset{O}{\overset{\Vert}{C}}{-}OH$	-	25

[a] Michel, G. and Renson, M., Spectrochim. Acta 23A, 1435 (1967).
k = scattering coefficient

[b] Freeman, S. K. Unpublished work.
I = Band peak height

REFERENCES

1. N. B. Colthup, L. H. Daly, and S. E. Wiberley, *Introduction to Infrared and Raman Spectroscopy* (Academic, New York, 1964).
2. K. Frei and H. H. Gunthard, J. Mol. Spectroscopy 5, 218 (1960).
3. G. Michel, Bull. Soc. Chim. Belges 68, 643 (1959).
4. S. K. Freeman and D. W. Mayo (unpublished work).
5. S. K. Freeman, Can. J. Spectroscopy 13, 99 (1968).
6. G. Allen and H. J. Bernstein, Can. J. Chem. 33, 1055 (1955).
7. G. Allen, P. S. Ellington, and G. D. Meakins, J. Chem. Soc. 1909 (1960).
8. C. L. Angell, P. J. Krueger, R. Lauzon, L. C. Leitch, K. Noack, R. D. Smith, and R. N. Jones, Spectochim. Acta 15, 926 (1959).
9. L. J. Bellamy and R. I. Williams, J. Chem. Soc. 4294 (1957).
10. K. Noack and R. N. Jones, Can. J. Chem. 39, 2225 (1961).
11. R. L. Erskine and E. S. Waight, J. Chem. Soc. 3425 (1960).
12. D. D. Faulk and A. Fry, J. Org. Chem. 35, 364 (1970).
13. K. Noack and R. N. Jones, Can. J. Chem. 39, 2201 (1961).
14. R. Mecke and K. Noack, Spectrochim. Acta 12, 391 (1958).
15. R. Mecke and K. Noack, Chem. Ber. 93, 210 (1960).
16. F. H. Cottee, B. P. Straughan, J. C. Timmons, W. F. Forbes, and R. Shilton, J. Chem. Soc. 1146 (1967).
17. G. Michel, Bull. Soc. Chim. Belges 68, 643 (1959).
18. T. L. Brown, J. Am. Chem. Soc. 80, 3513 (1958).

Chapter 5

ETHYLENIC DOUBLE BONDS

5.1 INTRODUCTION

To the best of the author's knowledge, no compound with a C=C moiety has failed to display a Raman band in the 1500-1900 cm^{-1} region. Similar to carbonyls (see p.77), stretching frequencies and, in some instances, depolarization ratios are excellent diagnostics for olefin type. The symmetry of an alkene is an important factor affecting the C=C stretching vibration in the IR. While the spectra of (Z) isomers generally have weak to moderate intensity absorption bands, (E) isomers show little or no absorptions due to C=C stretching (1). NMR interpretations sometimes tend to be equivocal for (E), (Z) isomerism of disubstituted double bonds, the presence of tetrasubstituted double bonds, and so on. Consequently, Raman spectroscopy is the definitive technique for determining ethylene groups. Isomerization of (E) or (Z) disubstituted olefins, cyclization of alkenes to cycloalkanes, and addition reactions such as hydrogenation, epoxidation, and chlorination can be monitored conveniently by measuring the decrease in Raman C=C band intensity (2). Complementary data obtained from IR, UV, and NMR often enable the investigator to gain a complete picture of C=C environments.

5.2 BAND LOCATIONS

5.2.1 Acyclic Alkenes

Utilizing =C-H and C=C stretching frequencies, little difficulty attends differentiating among most alkenes (Table 5.1).

TABLE 5.1. Characteristic Frequencies and Depolarization Ratios of Acyclic Alkenes

Ethylenic	Structure	=C-H Stretch (cm^{-1})	$=CH_2$[a] Stretch		C=C Stretch	
			(cm^{-1})	ρ	(cm^{-1})	ρ
Vinyl	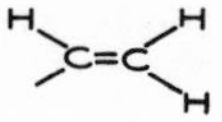 H(H)C=CH	3075-3100	3010-3025[b] 3075-3095[c]	0.01 0.5	1635-1650	0.04
Vinylidene	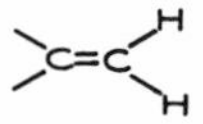 >C=C(H)H	--	3075-3100	--	1640-1660	0.04
(Z)[d]	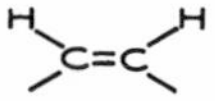 (H)C=C(H)	3000-3050	--	--	1635-1660	0.05
(E)[d]	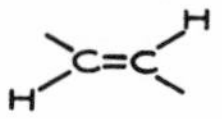 H-C=C-H	3000-3050	--	--	1665-1680	0.08
Trisubstituted	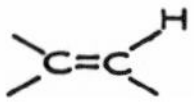 >C=C<H	2990-3050	--	--	1665-1695	0.1
Tetrasubstituted	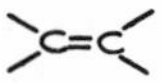 >C=C<	--	--	--	1665-1685[e]	0.1

[a]Also characterized by a 1410-1420 cm^{-1} deformation band

[b]Symmetric stretch

[c]Asymmetric stretch

[d]The terms "(Z)" and "(E)" are used by Chemical Abstracts, replacing "cis" and "trans" respectively.

[e]Bicyclobutylidene (◇=◇) -1746 cm^{-1}, ρ=0.2

Coupled with ancillary IR data, that is, the strong out-of-plane deformation bands associated with vinyl (near 910 and 990 cm^{-1}) and vinylidene (near 890 cm^{-1}) groups (1), it is possible to characterize all types of acyclic monoolefins with the doubly bonded carbons bearing hydrogen and/or carbon atoms. Substituting halogens (2,3) or other electronegative groups (2,4) usually lowers the frequency of the C=C stretching vibrations (Tables 5.2 and 5.3).

TABLE 5.2. Raman C=C Stretching Frequencies and Depolarization Values for Some Halogenated Alkenes

	X = CH_3	ρ	X = Cl	ρ	X = Br	ρ
$H_2C=CHX$	1648	--	1601	--	1593	--
$H_2C=CX_2$	1658	--	1611	--	--	--
XHC=CHX (H, H cis)	1670	0.03	1590	0.025	1587	0.03
XHC=CHX (H, H trans)	1684	--	1578	--	1582	--
$XHC=CX_2$	1682	--	1582	0.13	--	--
$X_2C=CX_2$	1676	--	1577	0.33	--	--
$XHC=CHCH_3$ (E) and (Z)	--	--	--	--	1642	0.02

The effect of phenyl substituents is shown in Table 5.4. Fluorinated alkenes have unusually high frequencies (1730-1800 cm^{-1}) (4).

Two or three bands, arising from conformational isomerism, generally appear in the Raman C=C stretching region for alkyl vinyl ethers (5,6). The observation of three bands implies the presence of at least two rotamers, for example, butyl vinyl ether (1612, 1643, 1655 cm^{-1}) exists in the liquid state as a mixture of three conformers (7), whereas two rotational isomers have been reported for methyl vinyl ether. In the latter instance, Raman (6), IR (8), and microwave (9) data indicate that the planar, or near planar, *cis* form I is the most stable.

C=C–O–C (cis)

I

TABLE 5.3. Effect of Electronegative Substituents on Raman C=C Stretching Frequencies and Depolarization Values

Chemical Name*	Structure	$\nu_{C=C}$ (cm^{-1})	ρ
(Z)-2-Heptene *5U2 -C*	$C{-}\overset{H}{C}{=}\overset{H}{C}{-}C{-}C_3$	1655	0.03
(Z)-Propenyl propyl disulfide *3SS1U2 -C*	$C{-}\overset{H}{C}{=}\overset{H}{C}{-}S{-}S{-}C_3$	1612	0.03
$\Delta^{2,2'}$Bi-1,3-dithiolane *T5SYSTJ B- 2U*	(ring structure: two 1,3-dithiolane rings joined by C=C)	1558	--
1-Pentene *4U1*	$C{=}C{-}C{-}C_2$	1644	0.03
Acrylonitrile *NC1U1*	$C{=}C{-}C{\equiv}N$	1610	0.025
Vinyl acetate *1VO1U1*	$C{=}C{-}O{-}\overset{O}{\overset{\parallel}{C}}{-}C$	1634	0.03
Allyl disulfide *1U2SS2U1*	$(C{=}C{-}C{-}S{-})_2$	1633	0.09
2-Methyl-2-butene *2UY*	$C{-}\underset{C}{C}{=}C{-}C$	1669	0.10
Ethyl (Z)-propenyl sulfide *2U1S2 -C*	$C_2{-}S{-}\overset{H}{C}{=}\overset{H}{C}{-}C$	1617	0.06
Ethyl (E)-propenyl sulfide *2U1S2 -T*	$C_2{-}S{-}\underset{H}{C}{=}\overset{H}{C}{-}C$	1638	0.08
Bis(2-methylpropenyl) disulfide *1Y&U1SS1UY*	$(C{-}\underset{C}{C}{=}C{-}S{-})_2$	1650	0.08
2-Methylpropenyl sulfide *1Y&U1S1UY*	$(C{-}\underset{C}{C}{=}C{-})_2\,S$	1648	0.08

*Wiswesser Line Notation in italics

TABLE 5.4. The Effect of Phenyl Substituents on Raman C=C Stretching Frequencies and Depolarization Values

Chemical Name*	Structure	$\nu_{C=C}$ (cm^{-1})	ρ
Phenyl ethylene *1U1R*	Φ-C=C	1634	-
2-Phenyl-1-propene *1UYR*	Φ-C(C)=C	1631	0.13
1,1-Diphenylethylene *1UYR&R*	$(\Phi)_2$-C=C	1610	0.12
(E)-1-Phenyl-1-propene *2U1R -C*	Φ-C(H)=C(H)-C	1668	0.18
(Z)-1,2-Diphenylethylene *R1U1R -T*	Φ-C(H)=C(H)-Φ	1629	0.21
(E)-1,2-Diphenylethylene *R1U1R -C*	Φ-C(H)=C(H)-Φ	1648	-
Tetrakis (2,4,6-trimethyl-phenyl) ethylene *1R C E D1 2U*	$R_2C=CR_2$ [a]	1520	-

*Wiswesser Line Notation in italics

[a] R = 2,4,6-trimethyl phenyl

5.2.2 Cyclic Alkenes

Many compounds of this class are liberally distributed in nature and the chemist frequently encounters them in synthetic work. The C=C stretching bands of exocyclic double bonds in unstrained ring systems have frequencies similar to acyclic vinylidene groups, but increased ring strain causes an increase in stretching frequencies (10) (Table 5.5). The double bond-single bond angle becomes larger with decreasing ring size and the steady increase in the C=C stretching frequency is due chiefly to an increasing interaction with the directly attached C-C bonds (11).

In contrast to the behavior of the vinylidene group in cyclic compounds, stretching frequencies of di- and trisubstituted endo double bonds decrease as ring strain increases (10,12) (Table 5.6).

5.3 BAND INTENSITIES AND BAND INTENSITY RATIOS

Acyclic and cyclic olefins below molecular weight of about 200 generally exhibit relatively strong C=C stretching Raman bands. In fact, confident assignments can be made in those instances where molecular weights reach approximately 500. The spectrum of cholesterol (m.wt. 386) appears in Figure 5.1. Armed with the knowledge that a compound of molecular weight of about 1000 contains one "unsaturation" (see p.144), it is not difficult to determine whether or not an ethylenic moiety is present. By contrast, analogous IR band intensities generally vary from zero to moderate for substances of molecular weight of about 200. The order of increasing intensity is (E)-disubstituted < (Z)-disubstituted < trisubstituted, vinyl, and vinylidene. Tetrasubstituted double bonds rarely are observed in IR spectra. Moreover, it is difficult to detect most types of IR C=C stretching bands in molecules containing an unconjugated carbonyl group because they usually are buried in the shoulder of the strong C=O absorption band.

Some unusual spectral properties are shown in compound II, containing a lone ethylenic group (13). It exhibits an unusually weak Raman band at 1690 cm^{-1} and an unexpectedly high wavelength UV maximum of 224 nm (ε = 5200; EtOH).

TABLE 5.5. Effect of Ring Strain on Exomethylene C=C Stretching Frequency

Methylene cyclo-	Structure	$\nu_{C=C}$ (cm^{-1})	ρ
Propane *L3YTJ AU1*		1780	-
Butane *L4YTJ AU1*		1676	0.11
Pentane *L5YTJ AU1*		1654	0.05
Hexane *L6YTJ AU1*	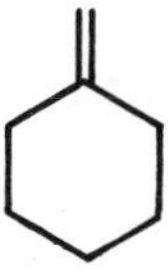	1650	0.08

An X-ray diffraction study revealed that the carbon-carbon double bond is highly twisted. This distortion from coplanarity reduces the π-electron overlap, decreasing the electron density between the ethylenic carbon atoms. The diminished polarizability is reflected by a relatively low intensity Raman band.

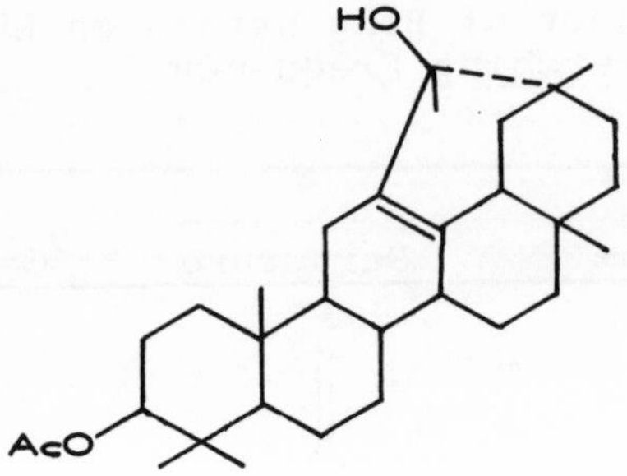

II

Contradictory reports appear in pre-laser Raman literature with regard to the dependence of band intensities on olefin type (14,16). Recent studies indicate a marked similarity in band strengths of various isolated double bonds occurring in the same or in different molecules (2,17) (Tables 5.7 and 5.8; Figure 5.2). Although additional confirmatory data are required before this relationship can be applied without reservation, it does allow for cautious interpretations. If the Raman spectrum of a purported homogeneous hydrocarbon, unconjugated carbonyl, and so on contains two C=C bands whose peak intensity ratio of high frequency to low frequency bands is <0.7 or >1.3, it is likely that the compound is impure. For example, a sample of commercial 2-pentene appeared to be homogeneous by NMR, mass spectrometry, and capillary column gas chromatography. An examination of the IR spectrum revealed the presence of an (E) double bond and a slight indication of a (Z) isomer. Two lines were clearly exhibited in the Raman effect at 1665 and 1650 cm^{-1} whose intensity ratio was 1.45, evidence of an about 70/30 mixture of (E) and (Z) isomeric forms (2). An interesting exception to this generalization is encountered in the terpene alcohol linalool (Figure 2.1) and its O-derivatives. Peak intensity and peak area ratios of the two ethylenic bands are 1.5 and 1.3, respectively. It has been suggested that the large difference in band intensities originates in a weak mechanical coupling of the C=C and C-O stretching vibrations (17).

Similar band intensities are displayed by the vinyl and trisubstituted double bonds of methyl isopimarate,

TABLE 5.6. Effect of Ring Strain and C=C Substitution on Raman Frequencies and Depolarization Ratios of Endocyclic Ethylenics

Chemical Name*	Structure	$\nu_{C=C}$ (cm^{-1})	ρ
Cyclopropene *L3 AHJ*		1641	-
2-Octyl cyclopropene-1-octanoic acid, methyl ester *L3 AHJ B8 C7VO1*	C_8 C_7COOC	1868	0.16
Tetrachloro cyclopropene *L3 AHJ-/G 4*	Cl Cl Cl Cl	1808	0.14
Cyclobutene *L4UTJ*		1566	-
1-Methyl cyclobutene *L4UTJ A*		1650	0.10
Cyclopentene *L5UTJ*		1612	∿0.01
2,3-Dihydrothiophene dioxide *T5SW BUTJ*	O_2 S	1618	0.02

a diterpene resin acid ester. However, when the ring C=C is located at carbon atoms 8 or 8(14), the vinyl/trisubstituted double bond intensity ratios are considerably lower (18) (Table 5.9). On the other hand, an equimolar mixture of methyl dihydropimarate and manoyl oxide (Table 5.7) shows two C=C stretching bands of nearly equal intensity.

TABLE 5.6. (Continued)

Chemical Name*	Structure	$\nu_{C=C}$ (cm^{-1})	ρ
2,3-Dihydrofuran *T5O BUTJ*		1622	0.02
2-Norbornene *L55 A CUTJ*		1572	0.03
3a,4,7,7a-Tetrahydro 4,7-methanoindene *L C555 A DU IU TJ*	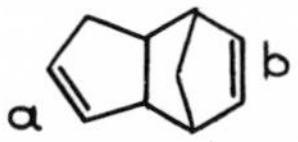 a, b	a-1616 b-1574	0.03 0.03
1-Methyl cyclopentene *L5UTJ A*	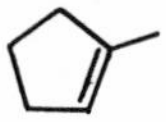	1655	∿0.03
2,3-Dihydro-5-methyl-furan *T5O BUTJ B*		1678	0.03
5-Methyl-3(2H)-furanone *T5O CV BHJ E*	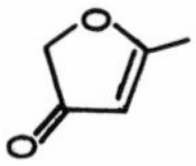	1679	0.03
1,2-Dimethyl cyclo-pentene *L5UTJ A B*	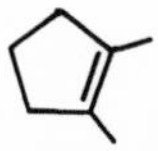	1689	0.05
2-Hydroxy-3-methyl-2-cyclopenten-1-one *L5V BUTJ BQ C*	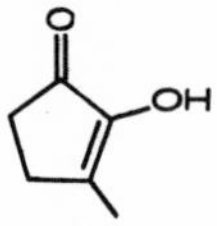 O, OH	1621	-
2,5-Dimethyl-3,4 (2H,5H) furandione *T5O CV BHJ DQ B E*	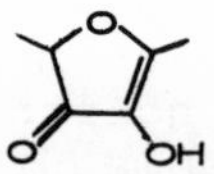 O, OH	1610	0.20

TABLE 5.6. (Continued)

Chemical Name*	Structure	$\nu_{C=C}$ (cm^{-1})	ρ
Cyclohexene *L6UTJ*		1650	0.02
3,4-Dihydro-2H-pyran *L6O BUTJ*		1650	0.02
1-Methyl cyclohexene *L6UTJ A*	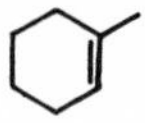	1674	0.09
1,1a,4,4a,5,6,7,8-Octahydro-2,4a,8,8-tetramethyl cyclopropa [d] naphthalene (Thujopsene) *L366 1A K AX DUTJ D G K K*	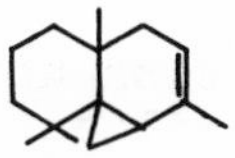	1680	0.09
2,3,4,7,8,8a-Hexahydro-1,5,8,8-tetramethyl 1H-3a,7-methanoazulene (Cedrene) *L B556 A 1UTJ E G G J*	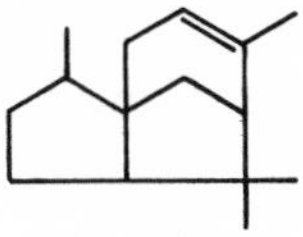	1663	0.09
2,7,7-Trimethyl,2-norbornene *L55 A CUTJ A A C*	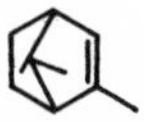	1657	0.10
3-Hydroxy-2-cyclohexen-1-one *L6V BUTJ CQ*	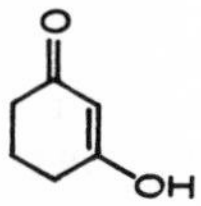	1650	0.17
1,2-Dimethyl cyclohexene *L6UTJ A B*	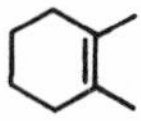	1680	0.10

*Wiswesser Line Notation in italics

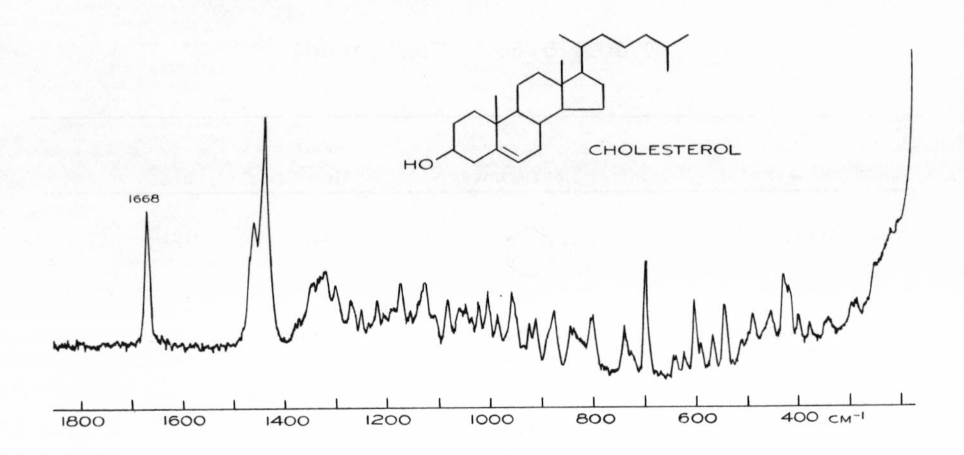

Figure 5.1. Raman spectrum of cholesterol.

Raman C=C intensity ratios have been employed to correct an error in structural assignments for dimethyl cyclopentadiene (commercial "methyl cyclopentadiene dimer") (2). Gas chromatography was reported to show that the dimer was a mixture of two compounds in approximately equal amounts and the NMR spectrum was interpreted to signify the presence of an equimolar mixture of III and IV (19).

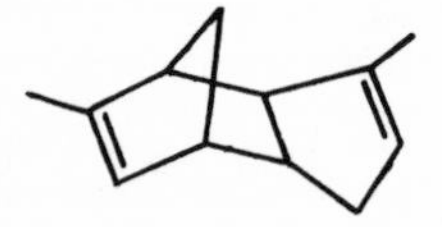

III IV

The Raman spectrum indicated a mixture of ∿75% V and ∿25% VI (Figure 5.3). Examination of the Raman spectra of the two GLC-isolated major components (30% and 43%), left no doubt that they were isomers of V only. Two minor constituents (9% and 15%) were found to be isomers of VI.

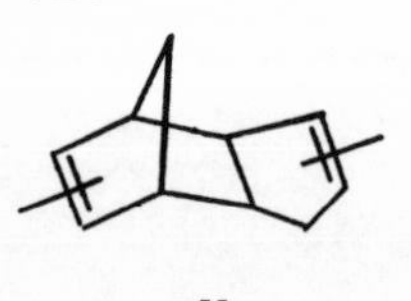

V

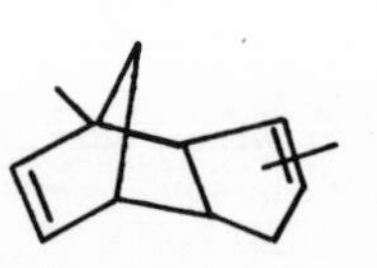

VI

TABLE 5.7. Intensity Ratios of Various Types of Ethylenic Double Bonds

Equimolor Mixture	C=C Intensity Ratio*
H H C–C=C–C–C–C H C–C=C–C–C–C H	0.9
C=C–C–C–C–C C H C–C=C–C–C–C H	1.0
$C=C-C_7$ $C-C=C-C_5$ C	1.0
OH $C=C-C_7-C-OH$	0.9
	0.85
	1.0
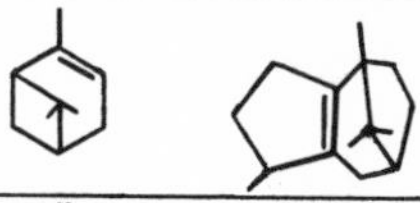	1.0
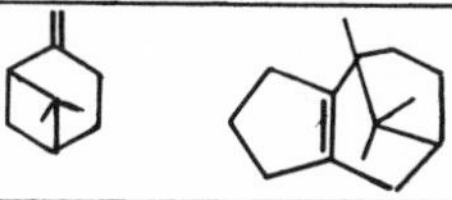	0.9
COOC	0.9

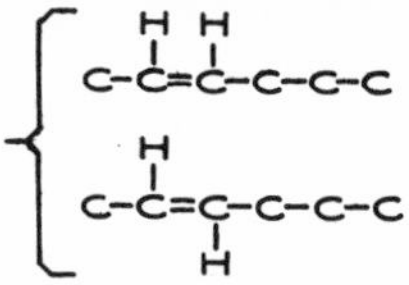

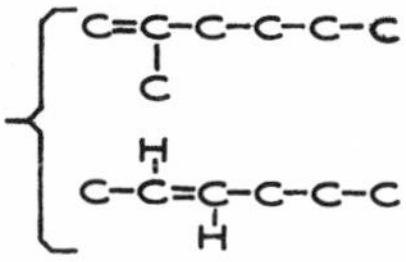

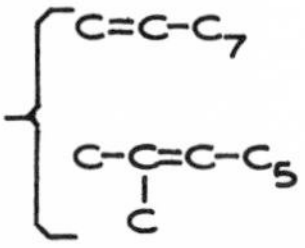

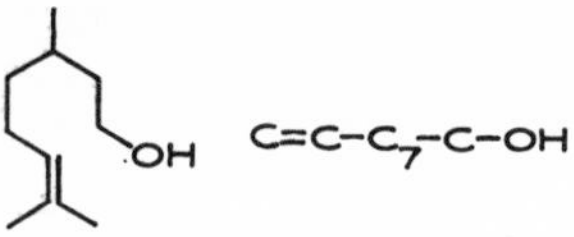

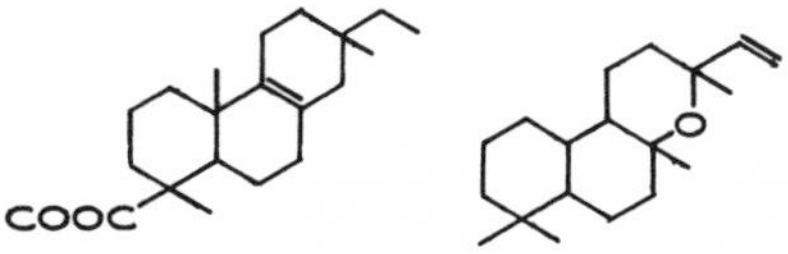

*I_{high}/I_{low} frequency bands

TABLE 5.8. Intensity Ratios of Various Types of Ethylenic Double Bonds in Unconjugated Dialkenes

Compound	C=C Intensity Ratio*	Compound	C=C Intensity Ratio*
	1.0		1.1
	1.0		1.0
	1.0		1.2
	1.0		
	1.0		0.75
	0.85		
	0.90		0.90
	0.80		

*I_{high}/I_{low} frequency bands

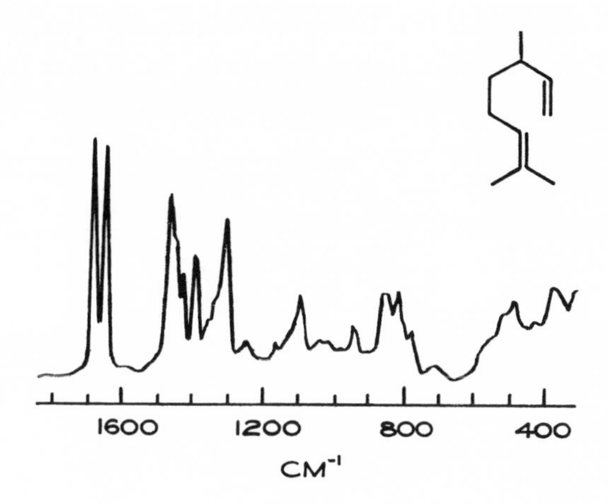

Figure 5.2. Raman spectrum of dihydromyrcene (17).

5.4 CONJUGATED ACYCLIC DIENES

Vibrational spectra of conjugated dienes usually possess two bands between 1500 and 1700 cm^{-1}. The complementarity of Raman and IR spectroscopy may be used to assign the C=C stretching vibrations by their relative intensities. The pertinent spectral region of myrcene, a terpene hydrocarbon, is shown in Figure 5.4. As a result of mechanical coupling in the conjugated system, the $C{=}CH_2$ stretching mode is split into two components (17):

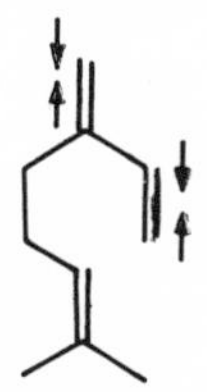

In-Phase
Symmetric Stretch
Raman and IR--
1640 cm^{-1}

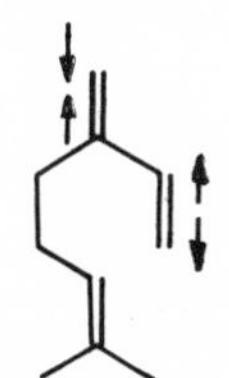

Out-of-Phase
Antisymmetric Stretch
IR--
1600 cm^{-1}

TABLE 5.9. Intensity Ratios of Ethylene Double Bonds in Resin Acid Esters

Chemical Name*	Structure	C=C Peak Intensity Ratio[a]
Methyl isopimarate *L B666 GUTJ A E1U1 E KVO1 K*	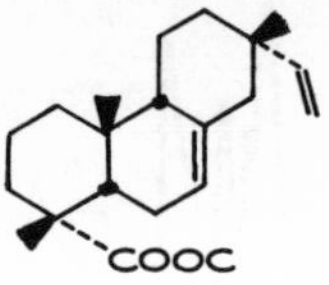	0.90
Methyl Δ^8 pimarate *L B666 BU- GTJ A E1U1 E KVO1 K*	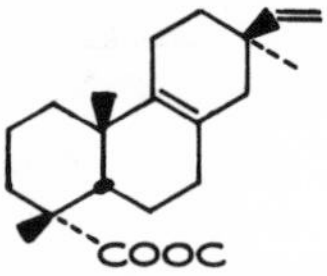	0.61
Methyl Δ^8 isopimarate *L B666 BU- GTJ A E1U1 E KVO1 K*	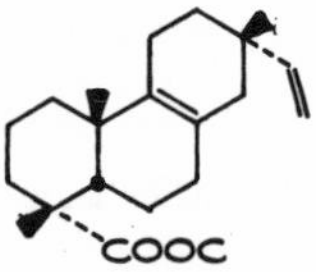	0.52
Methyl sandoracopimarate *L B666 FUTJ A E1U1 E KVO1 K*	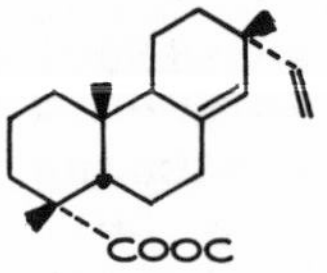	0.52
Methyl pimarate *L B666 FUTJ A E1U1 E KVO1 K*	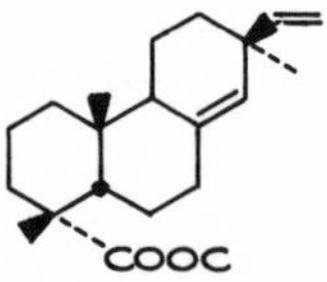	0.38

*Wiswesser Line Notation in italics

[a] I_{high}/I_{low} frequency bands

In the Raman effect, the in-phase symmetric stretch of the ethylenic double bond oscillators appears at 1640 cm^{-1}, considerably enhanced over the 1675 cm^{-1} trisubstituted C=C vibrational band. The out-of-phase antisymmetric stretching mode is particularly weak in this spectrum, but is clearly evident in the IR spectrum at 1600 cm^{-1}.

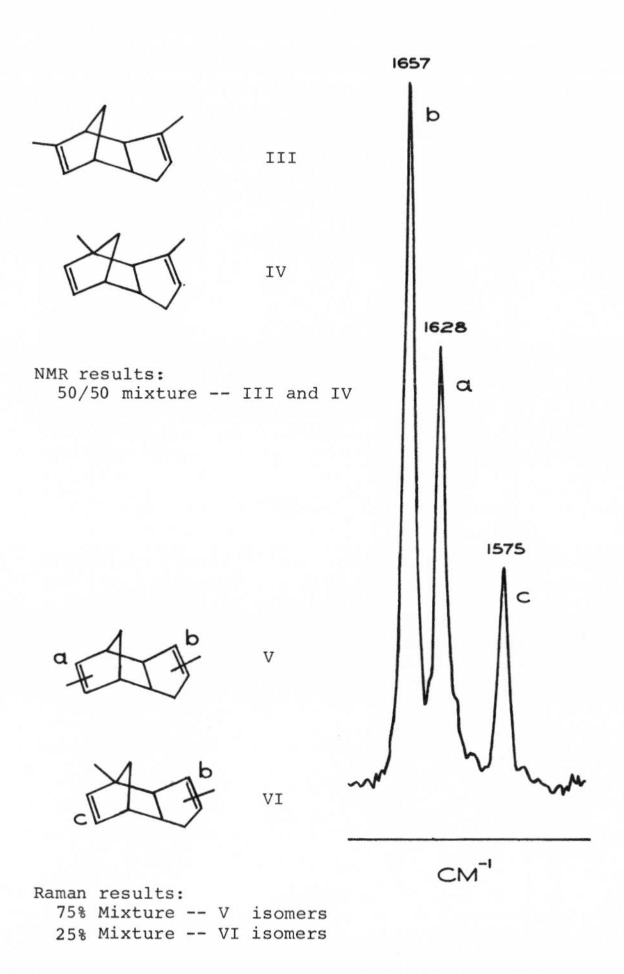

Figure 5.3. Composition of commercial dimethylcyclopentadiene.

Combined Raman and UV spectroscopic data can be employed to characterize conjugated dienes or trienes. The high UV extinction coefficients and intense Raman bands of such conjugated systems allow easy examination of a few μg sample (2) (Tables 5.10 and 5.11).

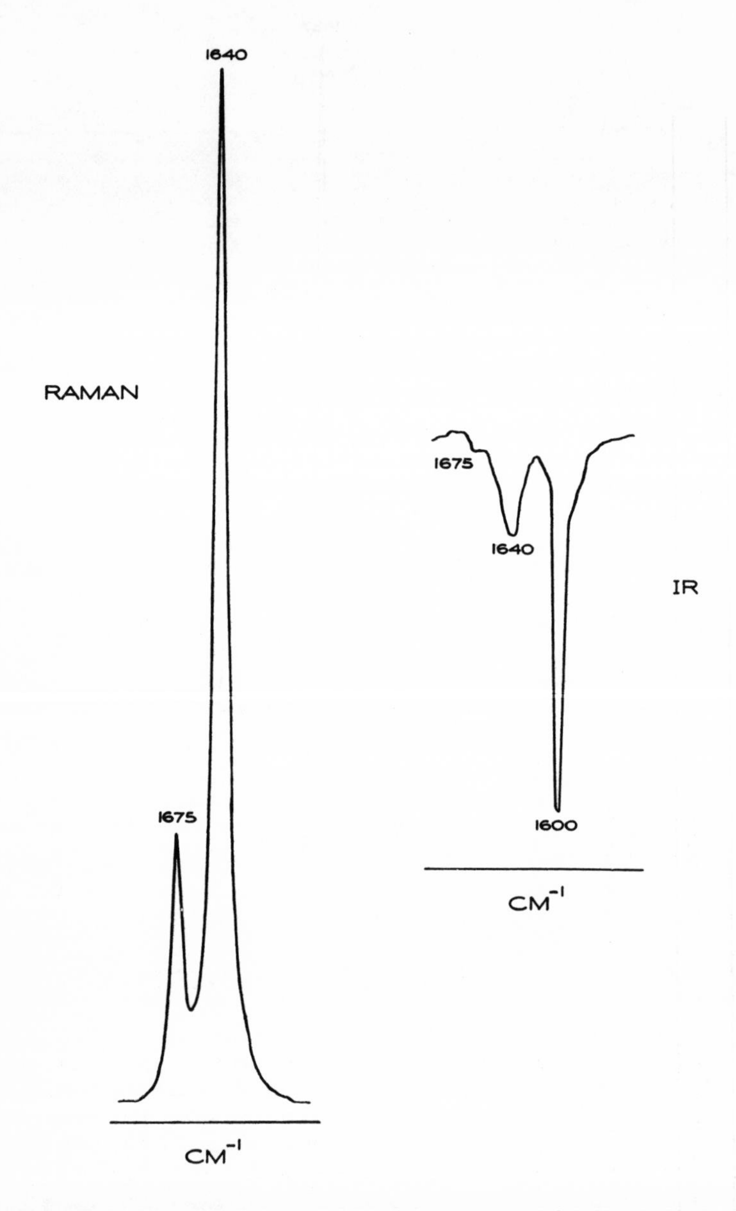

Figure 5.4. Myrcene. Mechanical coupling of the conjugated dienic system splits the $C{=}CH_2$ stretching mode into two components.

TABLE 5.10. Raman and UV Data for Some Acyclic Conjugated Dienes

Chemical Name*	Structure	$\nu_{C=C}$ (cm^{-1})	ρ	UV max (nm)	log ε[a]
1,(Z)-3-Pentadiene *2U2U1 -C*	C=C-C(H)=C(H)-C	1654	0.08	224	4.4
(Z)-2,(E)-4-Hexadiene *2U2U2 -TC*	C-C(H)=C(H)-C(H)=C(H)-C	1654	0.15	-	-
(E)-2,(E)-4-Hexadiene *2U2U2 -TT*	C-C(H)=C(H)-C(H)=C(H)-C	1667	0.18	236	4.35
3-Methyl-(Z)-2,(Z)-4-hexadiene *2UY1U2 -CC*	C-C=C-C(C)=C-C	1647	0.07	-	-
4-Methyl-1,3-pentadiene *1YU2U1*	C=C-C=C(C)-C	1654	0.08	233	-
2,3-Dimethyl-1,3-butadiene *1UY&YU1*	C=C(C)-C(C)=C	1634	0.08	-	-
2,5-Dimethyl-2,4-hexadiene *2UY&1UY*	C-C(C)=C-C(C)=C-C	1662	0.15	243	4.4

*Wiswesser Line Notation in italics

[a]EtOH solvent

TABLE 5.11. Raman and UV Data for Some Acyclic Conjugated Dienes

Chemical Name*	Structure	$\nu_{C=C}$ (cm^{-1})	C=C Peak Intensity[a]	ρ	UV max[a] (nm)	log ε[b]
Decene-1 *9U1*	C_8–C=C	1643	1	0.03	<210	-
3,7-Dimethyl-1,6-octadiene (Dihydromyrcene) *1Y&U3Y1U1*	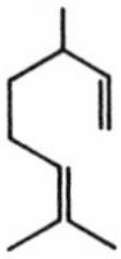	1645	1	0.04	<210	-
7-Methyl-3-methylene-1,6-octadiene (Myrcene) *1Y&U3YU1&1U1*		1644	5	0.10	225	4.3
3,7-Dimethyl-1,3,6-octatriene (Ocimene) *1Y&U2UY1U1*		1640	7	0.20	237	4.4
2,6-Dimethyl-2,4,6-octatriene (Alloocimene) *2UY&1U2UY*		1633	45	0.25	273	4.6

*Wiswesser Line Notation in italics

[a]Arbitrary units

[b]iso-Octane

5.5 CONJUGATED CYCLODIENES

Analogous to acyclic conjugated double bonds, their cyclic counterparts display intense Raman bands (Table 5.12). When ethylene groups are conjugated with a carbonyl, in addition to the C=O band there are observed as many bands between 1590 and 1670 cm^{-1} as there are C=C bonds.

TABLE 5.12. Raman and UV Data for Some Cyclic Conjugated Dienes

Chemical Name*	Structure	$\nu_{C=C}$ $(cm^{-1})^{a}$	ρ	UV max[b] (nm)	log ε^{b}
Cyclopentadiene *L5 AHJ*		1507	0.04	240 246 252 256 262	3.5 3.5 3.3 3.2 2.9
1-Methyl cyclopentadiene *L5 AHJ B*		1530	0.04	250	3.6
Hexachloro cyclopentadiene *L5 AHJ-/G 6*		1572	-	-	-
1,3-Cyclohexadiene *L6U CUTJ*		1575	0.05	257	3.2
1,3-Cyclooctadiene *L8U CUTJ*		1628	0.06	227	3.8

*Wiswesser Line Notation in italics

[a]Major band

[b]Isooctane solvent

These appear both in Raman and IR spectra, and the nature of the conjugated systems can be gained by relative band intensities. For example, one can differentiate between 1- and 4-unsaturated 3-ketosteroids as well as 1,4- and 4,6-dienic 3-ketosteroids (20) (Figure 5.5).

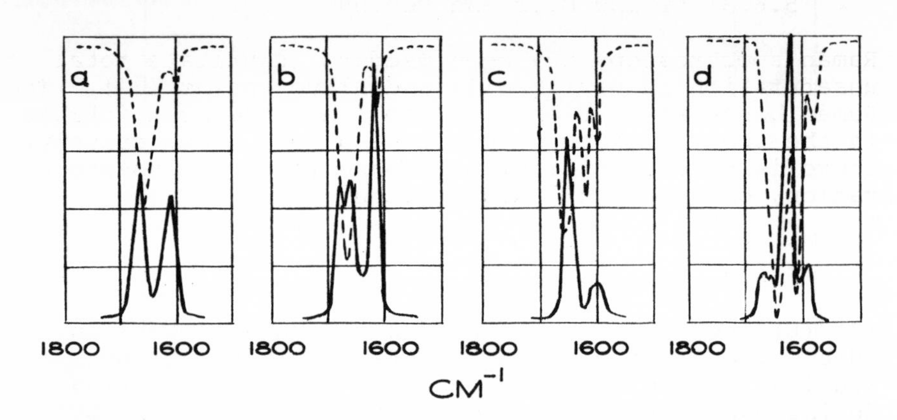

Figure 5.5. Partial Raman spectra of some ketosteroids (20). (a) 17 β-hydroxy androst-1-en-3-one; (b) 17 β-hydroxy androst-4-en-3-one (testosterone); (c) 17 β-hydroxy androsta-1,4-dien-3-one; (d) 17 β-hydroxy androsta-4,6-dien-3-one.

5.6 QUANTITATIVE ANALYSIS

5.6.1 Extent of Hydrogenation

Hydrogenation of alkenes (or cycloalkenes) to the corresponding alkanes (or cycloalkanes) can be monitored readily by Raman spectroscopy. For example, as little as about 0.05% dihydromyrcene can be determined in tetrahydromyrcene using a 200 mW Ar^+ laser (2).

5.6.2 Isomerization

In many instances, the degree of (Z) to (E) isomerization of disubstituted ethylenes can be ascertained by a ratio of the peak intensities of the about 1675 cm^{-1} (E) to about 1645 cm^{-1} (Z) bands and obtaining the percentage of product from a standard curve. The sensitivity of the procedure is dependent chiefly on the frequency difference (2).

5.6.3 Drying Oils and Resins

Raman spectroscopy has been used to quantitate total unsaturation in drying oils and alkyd resins (21). The Raman bands at 1638 and 1639 cm^{-1} are the most intense in the spectra of tung and linseed oils, respectively. There is no characteristic IR band for total unsaturation, only for (E) unsaturation.

5.6.4 Vegetable Oils

(Z)/(E) isomer composition of unsaturated esters in edible vegetable oils is of interest in nutrition studies. In addition, structural changes occurring during hydrogenation of the oils are accompanied by altered storage, flavor, and oxidative stability. Frequencies and intensities of (E) double bonds for several reference compounds are listed in Table 5.13 (22).

TABLE 5.13. Raman Frequencies of Reference Compounds

Compound	Class	$\nu_{C=C}$ stretch (cm^{-1})	(E) Intensity[a]
Methyl laurate	Saturated	--	--
Methyl oleate	(Z)	1655.1	0.078
Triolein	(Z)	1653.8	0.075
Methyl elaidate	(E)	1669.0	0.768
Trielaidin	(E)	1668.5	0.761
Methyl linoleate	(Z),(Z)	1656.1	0.149
Trilinolein	(Z),(Z)	1656.4	0.144
Methyl linoelaidate	(E),(E)	1671.5	0.826
Trilinoelaidin	(E),(E)	1670.5	0.821
Methyl linolenate	(Z),(Z),(Z)	1657.0	0.210
Trilinolenin	(Z),(Z),(Z)	1656.4	0.203

[a]Integrated area, corrected for background scattering at 1626 and 1691 cm^{-1}.

Raman and IR analyses for (E) isomer content of some vegetable oils appear in Table 5.14 (22).

TABLE 5.14. Comparison of Raman and IR Analyses (E) Isomer Content

Oil	Fatty Acids			Percent (E) Isomer	
	18:1	18:2	18:3	Raman	IR
High oleic safflower	80.7	12.2	--	0.0	0.0
Commercial safflower	12.8	77.9	--	0.0	0.0
Soybean 0	24.6	53.5	7.6	0.0	0.0
Soybean 1	25.9	53.2	6.0	1.5 + 0.59	1.6
Soybean 2	26.9	52.8	4.3	3.3 + 0.86	2.9
Soybean 4	33.6	50.0	1.2	6.4 + 0.84	7.9
Soybean 5	37.2	47.1	0.6	10.3 + 0.68	10.7
Soybean 7	45.0	39.8	--	14.5 + 0.80	16.4
Soybean 8	48.9	36.0	--	18.2 + 1.08	18.3
Soybean 9	58.2	26.2	--	23.0 + 0.36	23.9

A precision of about 1% can be obtained in (Z)/(E) isomer analysis of binary mixtures of methyl esters and triglycerides of monoenes and dienes and of hydrogenated vegetable oils.

Raman spectra of carbon tetrachloride solutions of the 15 isomeric methyl (Z)-octadecenoates and the 15 isomeric (E)-octadecenoic acids show that, when the double bond is isolated and not terminally located, the stretching vibration gives rise to a strong band at 1656 ± 1 cm^{-1} for the (Z) compounds and at 1670 ± 1 cm^{-1} for the (E) compounds (23).

5.6.5 Industrial Polymers

Ethylidenenorbornene (EN)

Analysis of EN in ethylene propylenediene-copolymer may be accomplished by an internal standard method (24).

The intensity ratio of the 1663 cm^{-1} C=C stretching band to the 1442 cm^{-1} methylene deformation band gives a linear relationship between 0% and 10% of EN.

Unsaturated Polyesters (25)

These resins can be easily cross-linked with various monomers to produce a range of polymer properties. During the polyesterification process a (Z) to (E) isomerization of maleic acid or maleic ester occurs which affects the structure and properties of the cross-linked resin. Raman spectroscopy offers a practical method of high sensitivity for studying the structure of cross-linked polyesters, even in the presence of fillers such as glass fibers or glass spheres. The Raman spectrum (1600-1800 cm^{-1}) of unsaturated polyester resins in different stages of the polyesterification reaction are presented in Figure 5.6. The intensity of the bands associated with the (E) isomer (1662 cm^{-1}) increases sharply from zero, and then gradually levels off. The higher-energy (Z) form is produced immediately after opening of the anhydride ring but isomerizes during the reaction.

A strong scattering at 1592 cm^{-1} arises from the C=C stretching mode of the cyclic anhydride starting material. The 1730 cm^{-1} carbonyl band is not affected by the polyesterification, and can be used as an internal standard band. After cross-linking, styrene copolymerizes preferentially with (E), or fumarate unsaturation.

5.6.6 Polybutadiene Rubbers (26)

Polymerization of 1,3-butadiene produces either 1,4 or 1,2 addition compounds (Figure 5.7). The 1,4 addition forms either the (Z) or (E) isomer, while the 1,2 addition proceeds either isotactically or syndiotactically. Based on a Raman spectral study of (Z)-1,4- (E)-1,4- and 1,2-polybutadienes, it appears that the Raman carbon-carbon double bond stretching vibrations (1650, 1664, and 1655 cm^{-1}, respectively) can be used to uniquely describe the structure content in polybutadienes.

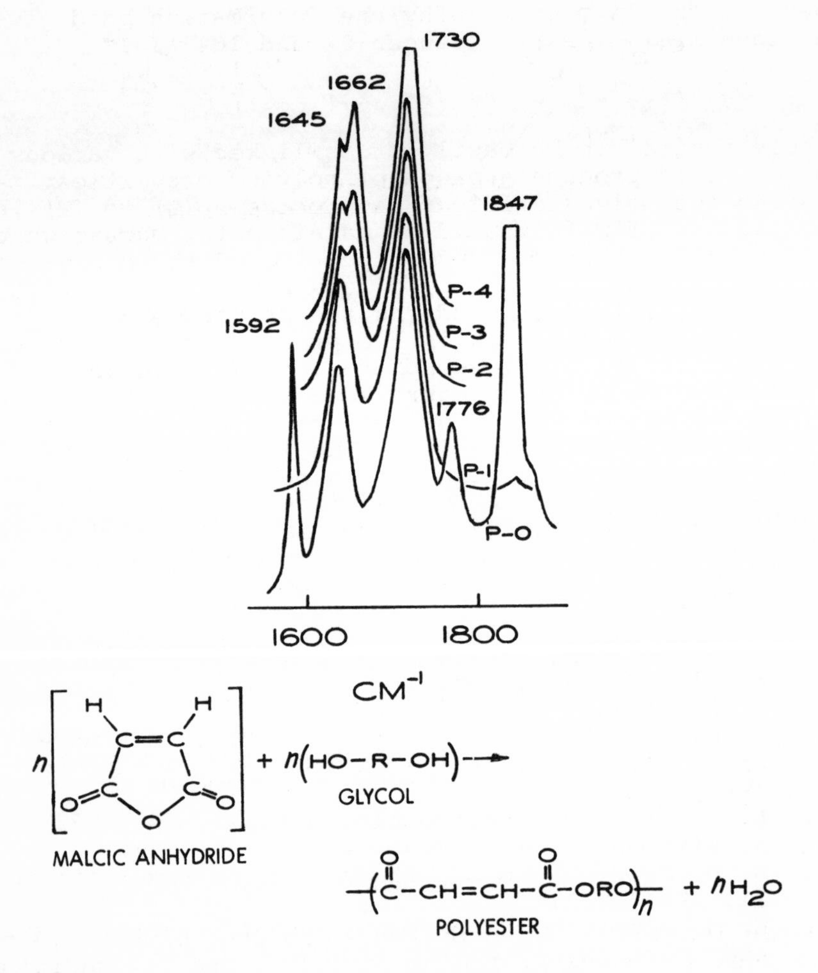

Figure 5.6. Raman spectra of the unsaturated polyester resins in five different stages (P = 0 – 4) of the polyesterification reaction (26).

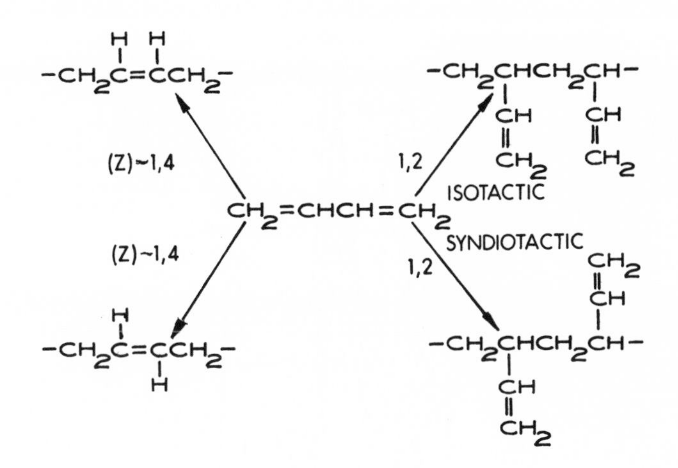

Figure 5.7. Polybutadiene polymerization routes.

REFERENCES

1. L. J. Bellamy, *The Infra-Red Spectra of Complex Molecules* (Wiley, New York, 1954).
2. S. K. Freeman (unpublished work).
3. K. W. F. Kohlrausch, in *Ramanspekten*, A. Eucken and K. L. Wolf (Eds.) (Edward Bros., Ann Arbor, 1945), Vol. 9.
4. L. J. Bellamy, Spectrochim. Acta 13, 60 (1958).
5. A. Kirrmann, Compt. Rend. 208, 353 (1939).
6. E. M. Popov and G. I. Kagan, Optics and Spectroscopy 11, 394 (1961).
7. E. M. Popov, N. S. Andreev, and G. I. Kagan, Optics and Spectroscopy 12, 17 (1962).
8. N. L. Owen and N. Sheppard, Proc. Chem. Soc., 264 (1963).
9. P. Cahill, L. P. Gold, and N. L. Owen, J. Chem. Phys. 48, 1620 (1968).
10. R. C. Lord and F. A. Miller, Appl. Spectrosc. 10, 115 (1956).
11. N. B. Colthup, J. Chem. Educ. 38, 394 (1961).
12. R. C. Lord and R. W. Walker, J. Am. Chem. Soc. 76, 2158 (1964).
13. W. E. Thiessen, H. A. Levy, W. G. Dauben, G. H. Beasley, and D. A. Cox, J. Am. Chem. Soc. 93, 4312 (1971).

14. J. J. Heigl, J. F. Black, and B. F. Dudenbostel, Anal. Chem. 21, 554 (1949).
15. H. Moser and U. Weber, Proc. Intern. Meeting Mol. Spectry., Bologna, Italy, 1959.
16. D. G. Rea, Anal. Chem. 32, 1638 (1960).
17. S. K. Freeman and D. W. Mayo, Appl. Spectrosc. 23, 610 (1969).
18. S. K. Freeman and D. Zinkel (unpublished work).
19. W. E. Franklin, J. Org. Chem. 35, 1794 (1970).
20. E. Steigner and B. Schrader, Liebigs Ann. Chem. 735, 15 (1970).
21. L. A. Neil and N. A. R. Falla, Chem. and Ind., 1349 (1971).
22. G. F. Bailey and R. J. Horvat, J. Am. Oil Chem. Soc. 49, 494 (1972).
23. J. E. Davies, P. Hodge, J. A. Barve, F. D. Gunstone, and I. A. Ismail, J. Chem. Soc., 1557 (1972).
24. G. Schreir and G. Peitscher, Z. Anal. Chem. 258, 199 (1972).
25. J. L. Koenig and P. T. K. Shih, J. Polymer Sci. A-2 10, 721 (1972).
26. S. W. Cornell and J. L. Koenig, Macromolecules 2, 539 (1969).

Chapter 6

RING SYSTEMS

6.1 ALICYCLICS

The past decade has seen a rapidly increasing number of naturally occurring substances isolated in limited quantities during the search for physiologically active materials as well as those displaying desirable flavor and odor properties. Molecular structure determination of such compounds are of major interest. The direct skeletal elucidation of these natural products, which often contain ring systems, has been restricted almost entirely to X-ray diffraction methods or chemical degradation. Because of their symmetry, vibrations associated with ring systems generally give rise to intense Raman emissions and weak IR absorptions. Incorporation of carbon rings into a molecular backbone increases the scattering intensities in the low frequency region when compared to the open chain analogs (Figure 6.1). Polar functional groups located on the ring scatter poorly in the Raman effect, so fingerprint assignments often allow confident identification of the molecular backbone. Several ring systems have already been characterized in this way.

6.1.1 Cyclohexanes

A strong, polarized band arising from a symmetric ring stretching, or breathing, vibration usually is found between 500 and 1200 cm^{-1} in the Raman spectra of ring systems. Detecting the often encountered cyclohexyl group by means of a frequency correlation pattern obviously would be of great interpretive importance. Unfortunately, Raman spectra of numerous cyclohexanes studied do not indicate such a pattern (1). However, of 40 alkyl cyclohexanes found in the American Petroleum

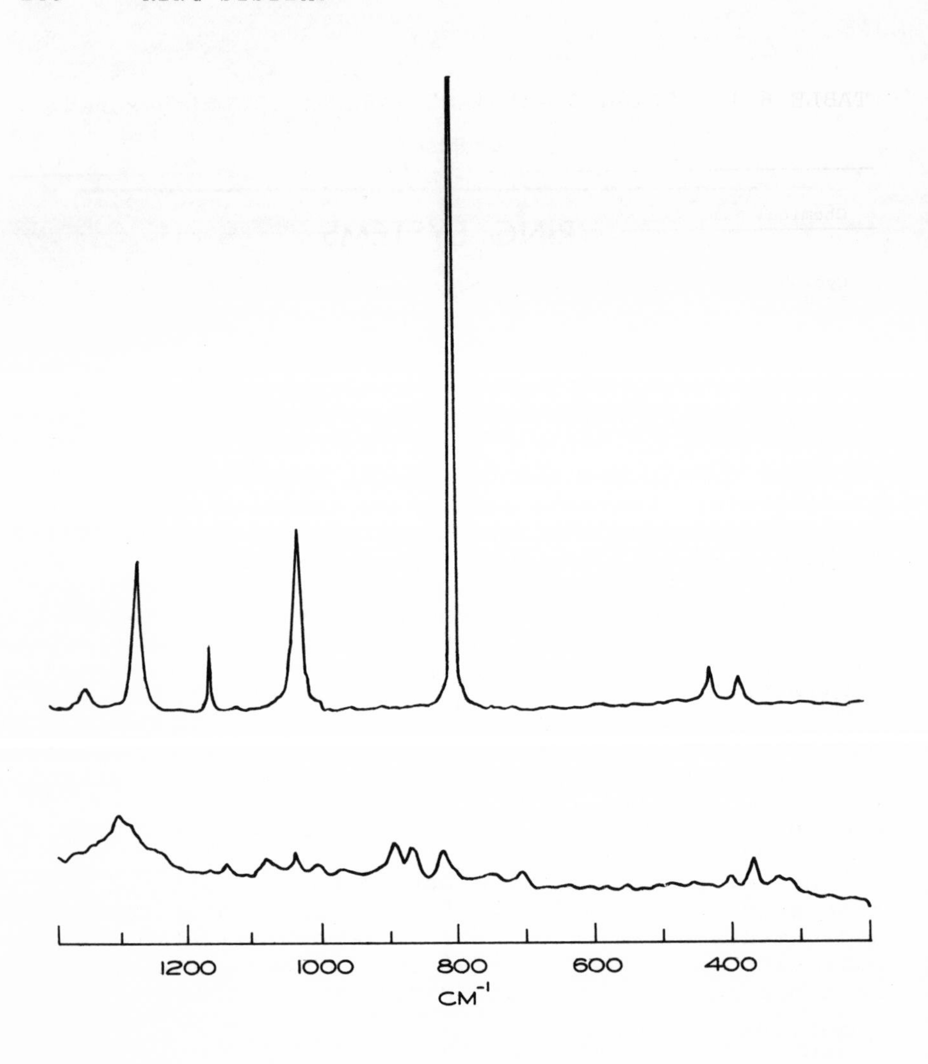

Figure 6.1. Raman spectra of cyclohexane (top) and hexane (bottom).

Institute's Raman spectral compilation (2) and reported in a Russian publication (3), 35 show an intense band between 700 and 800 cm^{-1}. The author's work (1) on compounds with C_6 rings indicates that the absence of a strong band in the 700-800 cm^{-1} range is evidence for the absence of an isolated cyclohexyl ring. Approximately 80% of the compounds examined exhibit a strong, polarized band in this region (Table 6.1).

TABLE 6.1. "Ring Breathing" Bands for Cyclohexanes

Chemical Name*	Structure	$\nu(cm^{-1})$	ρ	Rel. Int.[a]
Cyclohexane *L6TJ*		801	0.03	10
Cyclohexylamine *L6TJ AZ*	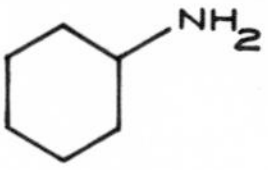 NH_2	782	0.05	10
1,2-Cyclohexyldiamine *L6TJ AZ BZ*	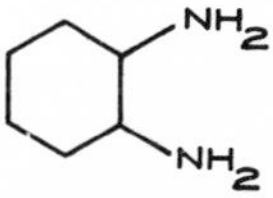 NH_2, NH_2	775	0.10	7
Cyclohexanol *L6TJ AQ*	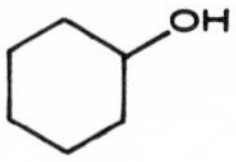 OH	742	0.04	10
p-Menthan-3-ol *L6TJ AY BQ D*	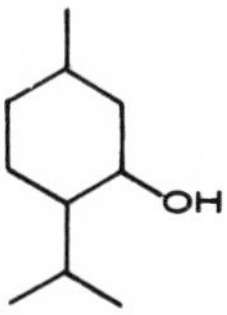 OH	780	-	10
Cyclohexane carboxylic acid *L6TJ AVQ*	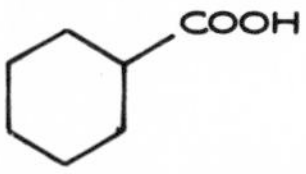 COOH	750	-	10
Methyl cyclohexane-carboxylate *L6TJ AVO1*	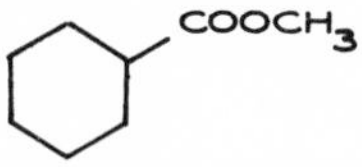 $COOCH_3$	755	0.04	10

TABLE 6.1. (Continued)

Chemical Name*	Structure	$\nu(cm^{-1})$	ρ	Rel. Int.[a]
p-Mentha-1,4-diene *L6U DUTJ AY D*	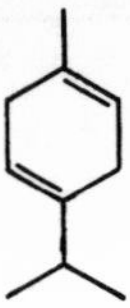	753	0.03	10
4-Isopropyl-1,4-cyclohexadiene-1-ethanol: formate *L6U DUTJ AY D2OVH*	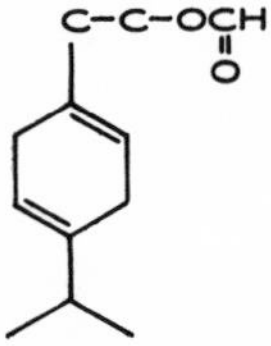 C-C-OCH=O	749	0.04	10
p-Mentha-1,3-diene *L6U CUTJ AY D*	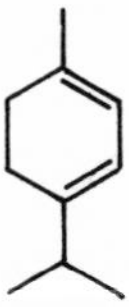	765	0.07	10
p-Menth-1-en-8-ol *L6UTJ A DXQ*	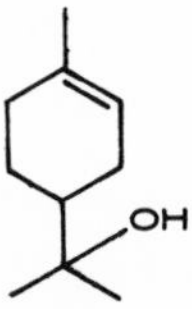 OH	756	0.02	10
p-Mentha-6,8-dien-2-ol *L6UTJ A DYU1 FQ*	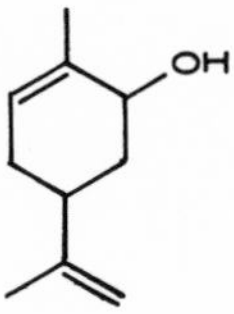 OH	772	0.05	10
1,3,5,5-Tetramethyl-1,3-cyclohexadiene *L6UTJ A C E E*	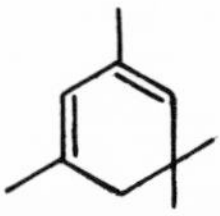	--	--	--

TABLE 6.1. (Continued)

Chemical Name*	Structure	ν(cm^{-1})	ρ	Rel. Int.[a]
p-Menth-8-en-2-ol *L6TJ A BQ DYU1*	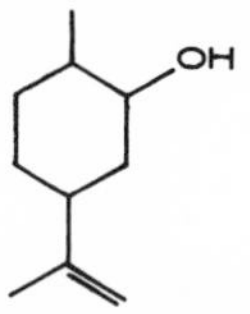 OH	755	0.04	10
p-Menth-1-en-4-ol *L6UTJ A DY DQ*	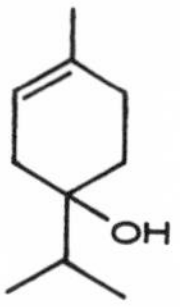 OH	732	0.02	10
p-Menth-8-en-1-ol *L6TJ AQ A DYU1*	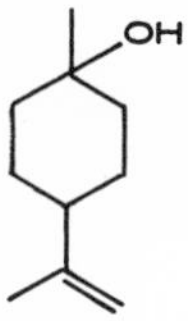 OH	730	0.015	10
γ,4-Dimethyl cyclohexane-propanol: acetate *L6TJ AY&2OV1 D*	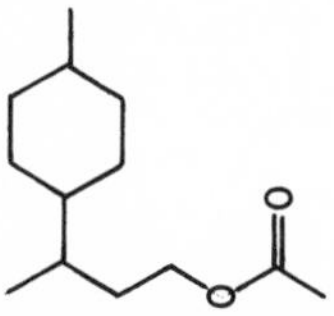 O, O	711	0.07	10
4-Isopropenyl-1-cyclo-hexene-1-carboxaldehyde *L6UTJ AVH DYU1*	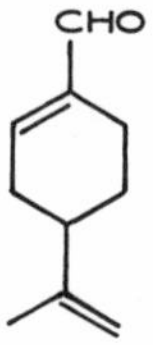 CHO	775	0.03	10

Many of the exceptions contain conjugated carbonyls. A considerable number of bi- and tricyclic materials (1) also show an intense band near 750 cm^{-1} (Table 6.2). High resolution mass spectroscopy is an excellent complement to the Raman effect in differentiating among cyclic, mono-, di-, and tricyclic compounds, thereby

TABLE 6.1. (Continued)

Chemical Name*	Structure	ν(cm^{-1})	ρ	Rel. Int.[a]
Cyclohexanethiol *L6TJ ASH*	SH	820	0.05	5
Cyclohexanone *L6VTJ*	O	749	0.03	10
2-Methyl cyclohexanone *L6VTJ B*	O	714	0.05	10
p-Menthan-3-one *L6VTJ BY E*	O	750	0.07	8
1,4-Cyclohexanedione *L6V DVTJ*	O O	727	--	10
p-Menth-3-yl carbazate *L6TJ AY BOVMZ D*	O-C(=O)-NHNH$_2$	780	0.06	10

confirming the presence or absence of a cyclohexyl ring system. The empirical formula, derived from the exact molecular weight, allows one to calculate the number of rings, double bonds, and triple bonds ("degree of unsaturation" or "index of hydrogen deficiency", Δ) present in a compound of formula $C_cH_hN_nO_{ox}$ by the simple expression

$$\Delta = c - h/2 + n/2 + 1$$

TABLE 6.1. (Continued)

Chemical Name*	Structure	ν(cm^{-1})	ρ	Rel. Int.[a]
2-Cyclohexene-1-one *L6V BUTJ*	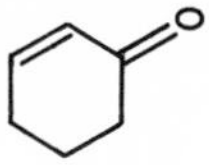	767	0.08	10
3-Methyl-2-cyclohexen-1-one *L6V BUTJ C*	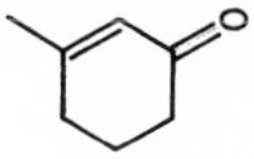	760	0.13	6
p-Mentha-6,8-dien-2-one *L6V BUTJ B EYU1*	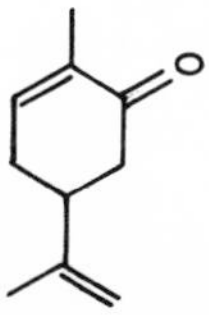	703	0.13	7
p-Menth-4(8)-en-3-one *L6VYTJ BUY E*	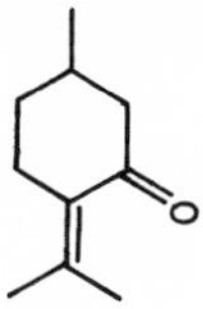	--	--	--
3,5,5-Trimethyl-2-cyclohexen-1-one *L6V BUTJ C E E*	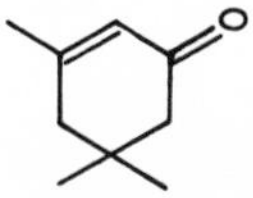	800	0.10	4
2-Hydroxy-2-cyclohexen-1-one *L6V BUTJ BQ*	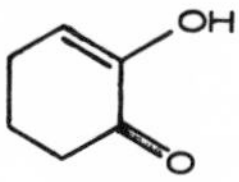	715	0.10	10
Methyl-3,4,6,6-tetramethyl-3-cyclohexen-1-yl ketone *L6UTJ A B D D EV1*	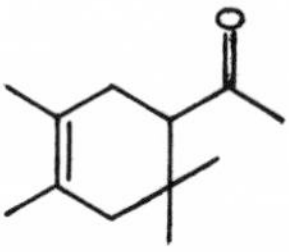	712	0.03	10

TABLE 6.1. (Continued)

Chemical Name*	Structure	ν(cm^{-1})	ρ	Rel. Int.[a]
Cyclohexylidene acetonitrile *L6YTJ AU1CN*	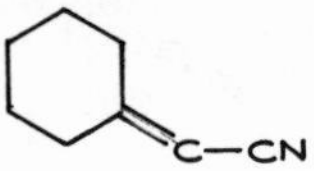 C—CN	780	0.08	10
Cyclohexene *L6UTJ*		822	0.015	10
3-Methyl cyclohexene *L6UTJ C*	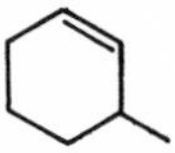	782	0.05	10
Methylene cyclohexane *L6YTJ AU1*	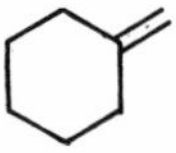	757	0.03	10
1-Methyl cyclohexene *L6UTJ A*	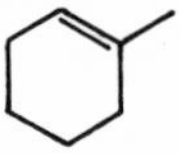	763	0.05	10
4-(4-Methyl-3-pentenyl)-3-cyclohexene-1-carboxaldehyde *L6UTJ A3UY DVH*	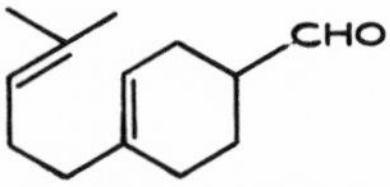 CHO	758	0.04	10
6-Methyl-4-(4-methyl-3-pentenyl)-3-cyclohexen-1-yl ketone *L6UTJ A3UY DV1 E*	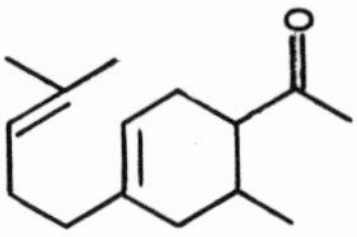 O	758	0.05	10

Molecular formulas also can be ascertained from low resolution mass spectra. The number of oxygen and nitrogen atoms usually can be gleaned from the molecular weight, mass spectral fragmentation pattern, Raman, IR, UV, and NMR spectra. Since the number of ethylenic and carbonylic double bonds as well as alkynes, nitriles, aromatic rings, and so on, generally can be determined

TABLE 6.1. (Continued)

Chemical Name*	Structure	ν(cm^{-1})	ρ	Rel. Int.[a]
2-Hydroxy-p-mentha-1,4 (8)-dien-3-one *L6VY EUTJ BUY E FQ*	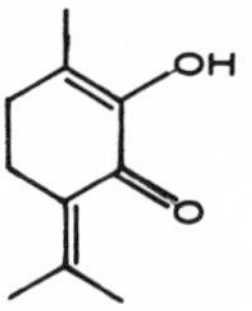	--	--	--
2-Chlorocyclohexanol *L6VTJ BG*	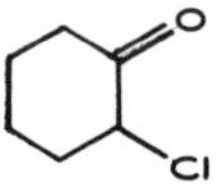	718	0.06	6
4-(2,6,6-Trimethyl-2-cyclohexen-1-yl)-3-buten-2-one *L6UTJ A E E F1U1V1*	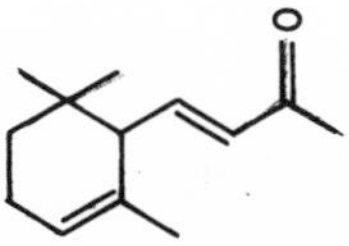	740	0.09	10
4-(2,6,6-Trimethyl-1-cyclohexen-1-yl)-3-buten-2-one *L6UTJ A B1U1V1 C C*	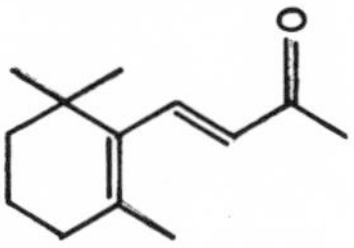	792	0.40	3

*Wiswesser Line Notation in italics

[a]The value of 10 signifies the strongest band between 300-1400 cm^{-1}. Numbers 9,8,7, etc. denote bands whose intensities are 90%, 80%, 70%, etc. of the strongest band.

from the Raman spectrum, with help from UV or NMR data if necessary, the number of alicyclic rings can be obtained by difference. In conjunction with the absence or presence of an intense, polarized band between 700 and 800 cm^{-1}, this type of information is quite helpful in deciding whether a cyclohexyl ring occurs in a particular molecule. It should be noted that strong Raman bands are observed near 750 cm^{-1} for cycloheptanes, cyclooctanes, cyclononanes (Table 6.3), and some cyclopentanes, although the majority of simple cyclopentanes examined (1) show a strong band between 800 and 900 cm^{-1} (Table 6.4).

TABLE 6.2. Ring Breathing Bands of Some Bicyclic Compounds[a]

Chemical Name*	Structure	$\nu(cm^{-1})$ [a]	ρ
1,2-Epoxy cyclohexane *T36 BOTJ*	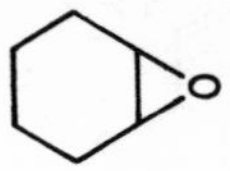	781	0.03
4,8-Epoxy-p-menth-1-ene *T3OX CHJ C C B-& AL6X_ CUTJ D*	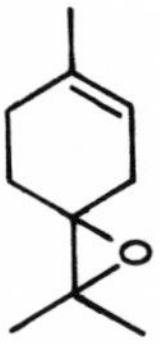	738	0.04
2-Carene *L36 DUTJ B B E*	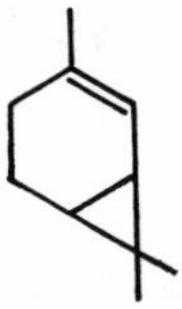	720	0.07
2-Caren-4-yl methyl ketone *L36 DUTJ B B E FV1*	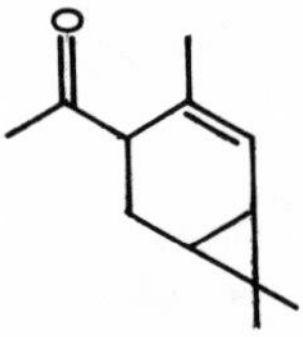	733	0.03
4(10)-Thujene *L35 DYTJ AY DU1*	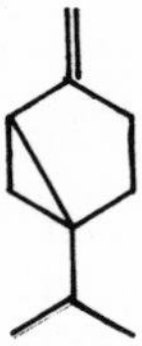	787	0.05
Borneol *L55 ATJ A A B CQ*	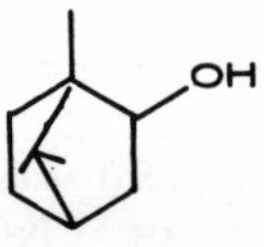	852	-

TABLE 6.2. (Continued)

Chemical Name*	Structure	$\nu(cm^{-1})^a$	ρ
Pinane *L46 ATJ A A E*	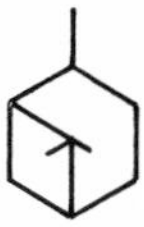	662	0.04
Methyl 1-isopropyl-4-methylbicyclo[2.2.2]oct-5-ene-2(or 3)-carboxylate *L67 A B AUTJ CY F G GVO1*	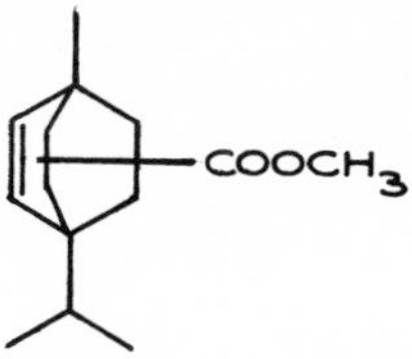 $COOCH_3$	727	0.02
1,8-Epoxy-p-menthane *T66 A B AOTJ B B F*	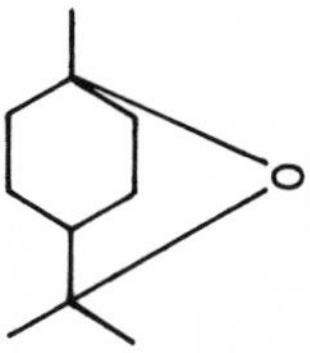 O	653	0.03
trans Hexahydroindan *L56TJ -T*	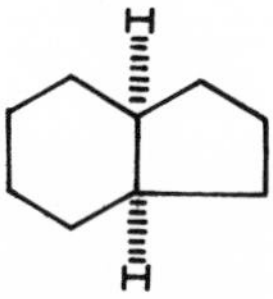	765	0.02
cis Decahydronaphthalene *L66TJ*	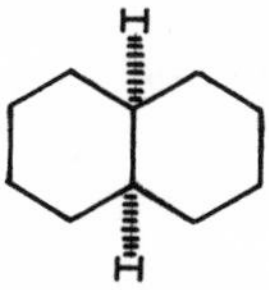	755	0.04

*Wiswesser Line Notation in italics

[a]All bands except the one for thujene (rel. int. 8) are the most intense (rel. int. 10) between 300 and 1400 cm^{-1}.

TABLE 6.3. Most Intense Bands of Some Cycloheptanes, Cyclooctanes, and Cyclononanes

Chemical Name*	Structure	$\nu(cm^{-1})$ [a]	ρ
Cycloheptane *L7TJ*		732	0.025
Cycloheptanol *L7TJ AQ*	OH	728	0.04
Methyl cycloheptane *L7TJ A*	C	710 722	0.03
Cyclooctane *L8TJ*		700	0.025
Cyclooctene *L8UTJ*		702	0.03
Cyclononanone *L9VTJ*	C_9, O	700	0.02

*Wiswesser Line Notation in italics

[a]Strongest band between 300-1400 cm^{-1}

6.1.2 Cedranes

A skeletal frequency correlation for cedrane-type molecules, sesquiterpenes with a closely knit tricyclic system, has been reported (4). The relatively complex spectrum of cedrane is shown in Figure 6.2.

TABLE 6.4. Ring Breathing Bands of Some Cyclopentanes

Chemical Name*	Structure	$\nu(cm^{-1})$[a]	ρ
Cyclopentane *L5TJ*		891	0.025
Methyl cyclopentane *L5TJ A*	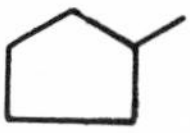	892	0.03
Cyclopentene *L5UTJ*		900	0.04
1-Methyl cyclopentene *L5UTJ A*	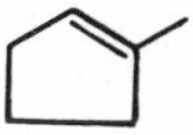	894	0.04
1,2-Dimethyl cyclo-pentene *L5UTJ A B*	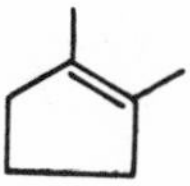	913	0.04
Methylene cyclopentane *L5YTJ AU1*		895	0.04
Cyclopentanone *L5VTJ*		892	0.025
3-Methyl-2-cyclopenten-1-one *L5V BUTJ C*	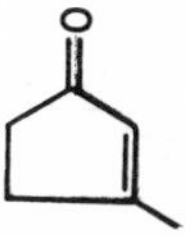	848	-
1,1-Dimethyl cyclopentane *L5TJ A A*		768	0.04

*Wiswesser Line Notation in italics

[a]All bands except the one for 3-methyl-2-cyclopenten-1-one (rel. int. 5) are the most intense (rel. int. 10) between 1300 and 1400 cm^{-1}).

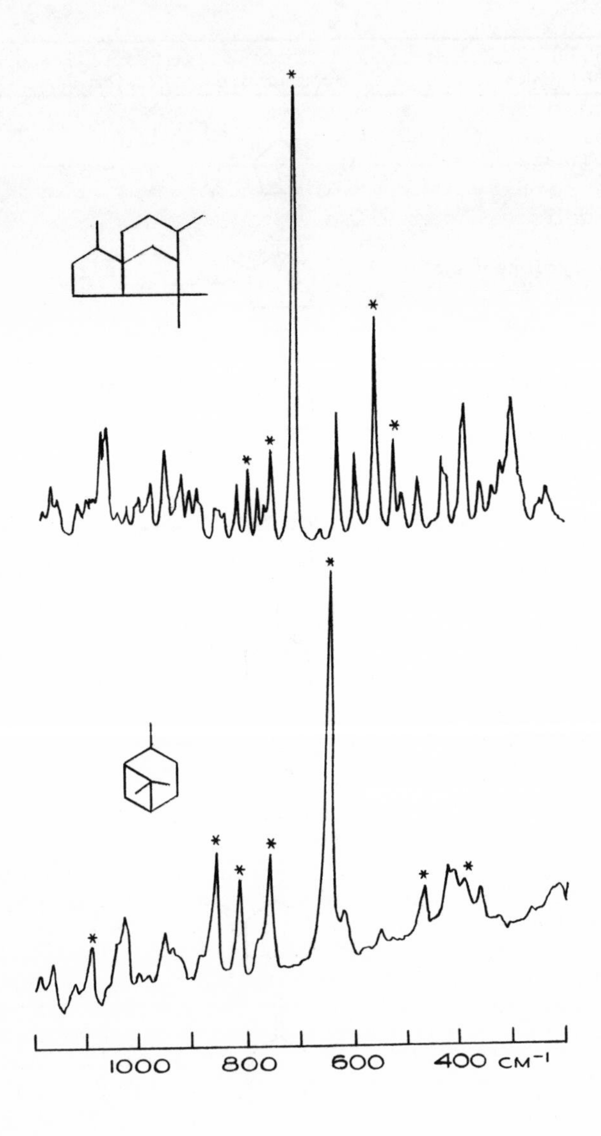

Figure 6.2. Raman spectra of cedrane (top) and *cis* pinane (bottom) (4).

For a group of cedranes examined (Table 6.5), five of the seven most intense bands observed between 500 and 900 cm^{-1} appear in a series of relatively narrow wave-number ranges correlatable with the cedrane skeleton

(Table 6.6). A band of variable intensity at about 320 cm^{-1}, characterized by a ρ value of 0.2-0.4, may be included in the correlation pattern. Of the many mono-, bi- and tricyclic naturally occurring products spectrally examined to test the validity of the cedrane skeletal pattern, none was encountered that successfully met the cedrane ring system constraints.

6.1.3 Pinanes

The number of bands observed in a vibrational spectrum increases with molecular complexity. Consequently, pinanes display simpler profiles than cedranes and other unsymmetrical tricyclic sesquiterpenes in the 300-1000 cm^{-1} Raman region (Figure 6.2). The spectra of pinane derivatives reveal the presence of seven common bands between 375 and 1100 cm^{-1} (5) (Tables 6.7 and 6.8). One of the strongest bands is highly polarized ($\rho \sim 0.05$) and appears at about 650 cm^{-1}. Spectral patterns of other bicyclo C_{10} ring systems (Table 6.9) and of many cyclopentanes, cyclohexanes, and cycloheptanes fail to meet the seven-band correlation established for the pinane skeleton.

6.1.4 Barbiturates (Pyrimidines)

With regard to the identification of the widely used barbiturates, Raman spectroscopy offers advantages over such methods as IR, UV, and thin-layer chromatography in the ease of sample handling and ability to work in aqueous solutions. These techniques usually involve separations and laborious sample preparation. The pyrimidine ring, present in all barbituric acids (Table 6.10) and their salts, can be identified using characteristic Raman carbonyl stretching and ring vibrations (6). In addition to at least two carbonyl bands and an intense ring breathing band, groups of bands are common to the barbiturates (Tables 6.10 and 6.11) (6). The Raman spectrum of phenobarbital appears in Figure 6.3 (7).

TABLE 6.5. Cedranes Characterized by Raman Skeletal Vibrations (4)

Trivial Name	Chemical Name*	Structure
α-Cedrene	Cedr-8-ene *L65 B5 A 1B BX DUTJ E G_ G K*	
β-Cedrene	Cedr-8(15)-ene *L65 B5 A 1B BX EYTJ EU1_ G G K*	
-	8βH-Cedrane,8,9-epoxy *T D365 B5 A 1B BX EOTJ_ F H H L*	
-	8βH-Cedrane,8,13-epoxy *T655 C5 B 2AC CX GOTJ_ F I M*	
Cedrenyl acetate	8βH-Cedran-8-ol: acetate *L65 B5 A 1B BXTJ EOV1 E_ G G K*	
Cedrone	8βH-Cedran-9-one *L65 B5 A 1B BX DVTJ E_ G G K*	
β-Isobiotol	Cedr-8(15)-en-3-ol *L65 B5 A 1B BX EYTJ EU1_ G G JQ K*	
β-Biotol	Cedr-8(15)-en-4-ol *L65 B5 A 1B BX EYTJ EU1_ G G JQ K*	

TABLE 6.5. (Continued)

Cedrol	8βH-Cedran-8-ol *L65 B5 A 1B BXTJ EQ E_ G G K*	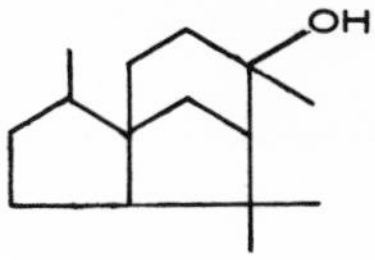 OH
Cedrenone	Cedr-8-en-10-one *L65 B5 A 1B BX CV DUTJ_ E G G K*	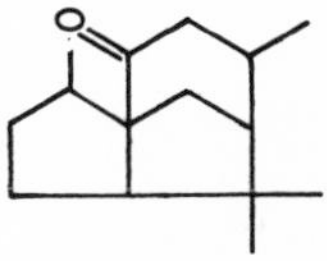 O
Cedrenal	Cedr-8-en-15-al *L65 B5 A 1B BX DUTJ_ EVH G G K*	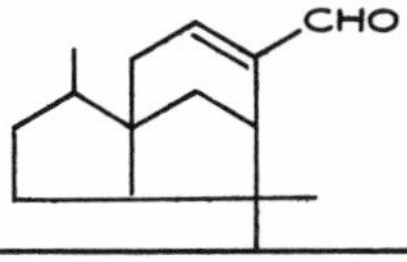 CHO

*Wiswesser Line Notation in italics

TABLE 6.6. Five-Band Correlation for the Cedrane Ring with Intensity Constraint[a] (4)

812 ± 10 cm^{-1}
775 ± 15
730 ± 15
582 ± 20
534 ± 15

[a]Band must be one of seven most intense bands in the 900 to 500 cm^{-1} region.

6.1.5 Hypoxanthines

The spectra of inosine (hypoxanthine), inosine-5'-monophosphate, and 1-methylinosine have been recorded in acidic, neutral, and basic H_2O and D_2O solutions (8).

TABLE 6.7. Pinanes Characterized by Raman Skeletal Vibrations (5)

Trivial Name	Chemical Name*	Structure
cis Pinane	cis Pinane *L46 ATJ A A E -C*	
trans Pinane	trans Pinane *L46 ATJ A A E -T*	
β-Pinene	2(10)-Pinene *L46 A EYTJ A A EU1*	
α-Pinene	2-Pinene *L46 A EUTJ A A E*	
Nopadiene	6,6-Dimethyl-2-vinyl-2-norpinene *L46 A EUTJ A A E1U1*	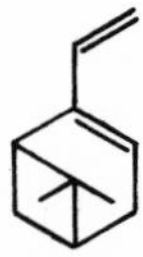
Myrtenol	2-Pinen-2-ol *L46 A EUTJ A A E1Q*	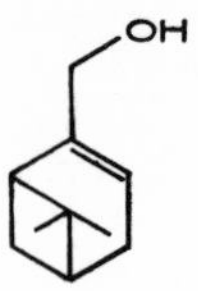 OH
Myrtenal	2-Pinen-2-al *L46 A EUTJ A A EVH*	 CHO
Pinoacetaldehyde	2-Pinene-10-acetaldehyde *L46 A EUTJ A A E1VH*	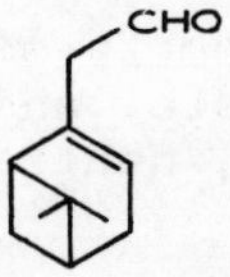 CHO

TABLE 6.7. (Continued)

Nopol	2-Pinene-2-ethanol *L46 A EUTJ A A E2Q*	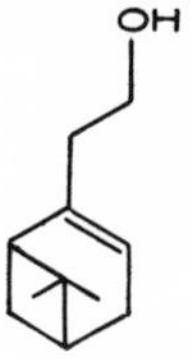
Pinocarveol	2-(10)-Pinen-3-ol *L46 A EUTJ A A EU1 FQ*	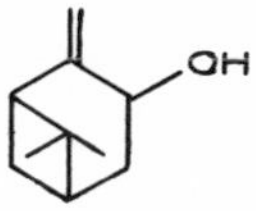
Isopinocamphone	3-Pinanone *L46 A FVTJ A A E*	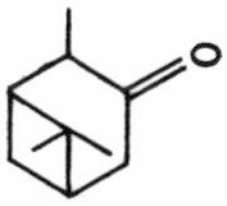
Verbenone	2-Pinen-4-one *L46 A FV EUTJ A A E*	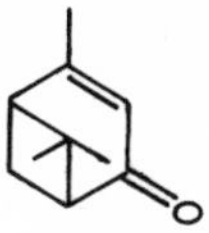
Myrtenyl chloride	10-Chloro-2-pinene *L46 A EUTJ A A E1G*	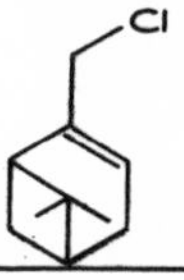

*Wiswesser Line Notation in italics

Certain bands of all derivatives are not appreciably changed in frequency or intensity and therefore can be assigned to skeletal vibrations (Table 6.12). This type of information is valuable in correlating specific spectral changes in molecular structure and for characterizing the hypoxanthine ring system in naturally occurring biopolymers.

6.1.6 Isochroman Musks

The Raman effect has been employed successfully as a probe for detecting conformational isomers in a variety of organic compounds (see, e.g., References 9-12).

TABLE 6.8. Seven-Band Correlation for the Pinane Ring

1082 ± 3 cm^{-1}[a]
852 ± 13
822 ± 4
758 ± 17
657 ± 26[b]
472 ± 12
387 ± 12[a]

[a]Usually weak intensity.

[b]Most intense for ten of the twelve substances examined; second most intense for myrtenal and myrtenol.

Generally, the Raman or IR spectrum of a nonlinear molecule comprising N atoms cannot contain more than $3N-6$ fundamental bands. However, for a mixture of conformers the number of bands exceeds the theoretical value for a single molecular form. Vibrational spectroscopic evidence for rotational isomerism may be obtained by comparison of spectra generated on two physical states of a substance. If the vibrational spectra of a liquid and solid substance are essentially identical, the material may be assumed to exist as a single conformation in both states. However, diminution in intensity or disappearance of specific bands on going from the liquid or solid states is good evidence for the presence of more than one rotamer in the liquid. The higher-energy (less stable) conformer(s) generally does not persist in the solid; hence those bands that are absent in the solid phase are attributed to the more energetic form. An interesting insight into an odor-molecular structure correlation has been gained by applying this general rule (1,13).

TABLE 6.9. Bicyclo C-10 Compounds Failing to Meet the Pinane Seven-Band Correlation Scheme (5)

Trivial Name	Chemical Name*	Structure
Camphene	Camphene *L55 A CYTJ CU1 D D*	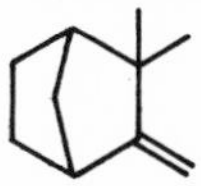
Borneol and iso-Borneol	Borneol and iso-Borneol *L55 ATJ A A B CQ*	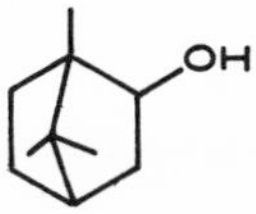 OH
3-Carene	3-Carene *L36 EUTJ B B E*	
Sabinene	4(10)-Thujene *L35 DYTJ AY DU1*	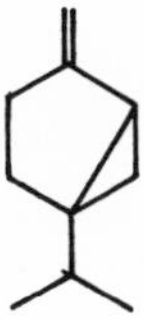
Sabinol	4(10)-Thujene-3-ol *L35 DYTJ AY DU1 EQ*	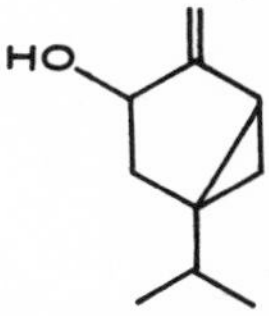 HO
Pinol	6,8-Epoxy-p-menth-1-ene *T56 A CO GUTJ D D H*	 O
1,8-Cineol	1,8-Epoxy-p-menthen *T66 A B AOTJ B B*	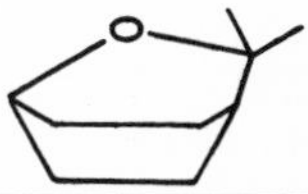 O

*Wiswesser Line Notation in italics

TABLE 6.10. Some Common Barbiturates

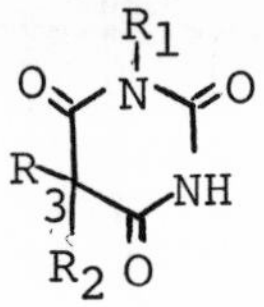

Barbituric Acid	R_1	R_2	R_3
Barbital	H	Ethyl	Ethyl
Phenobarbital	H	Phenyl	Ethyl
Pentobarbital	H	Ethyl	1-Methylbutyl
Secobarbital	H	Allyl	2-Methylbutyl
Amobarbital	H	Ethyl	4-Methylbutyl
Mephobarbital	Methyl	Ethyl	Phenyl
Hexobarbital	Methyl	Ethyl	1-Cyclohexenyl

TABLE 6.11. Characteristic Raman Bands for Barbituric Acids and Their Sodium Salts (6)

Barbituric Acids		Sodium Salts of Barbituric Acids			
$\nu(cm^{-1})$	Rel. Int.	$\nu(cm^{-1})$	Rel. Int.	$\nu(cm^{-1})$	Rel. Int.
416 ± 11	w-m	338 ± 4	w	1083 ± 10	m
599 ± 16	w-m	414 ± 4	w-m	1311 ± 4	w
629 ± 8	s	442 ± 3	w-m	1343 ± 5	m
1053 ± 7	w	522 ± 2	m	1446 ± 5	m-w
1078 ± 10	w-m	652 ± 4	s	1461 ± 1	m-s
1152 ± 7	w-m	696 ± 5	w	1585 ± 15	w
1321 ± 9	w	753 ± 3	w	1657 ± 3	m-s
1447 ± 5	m	788 ± 3	vw	1693 ± 3	w-s
1692 ± 6	m-s	955 ± 5	w-m		
1737 ± 8	m-vs				

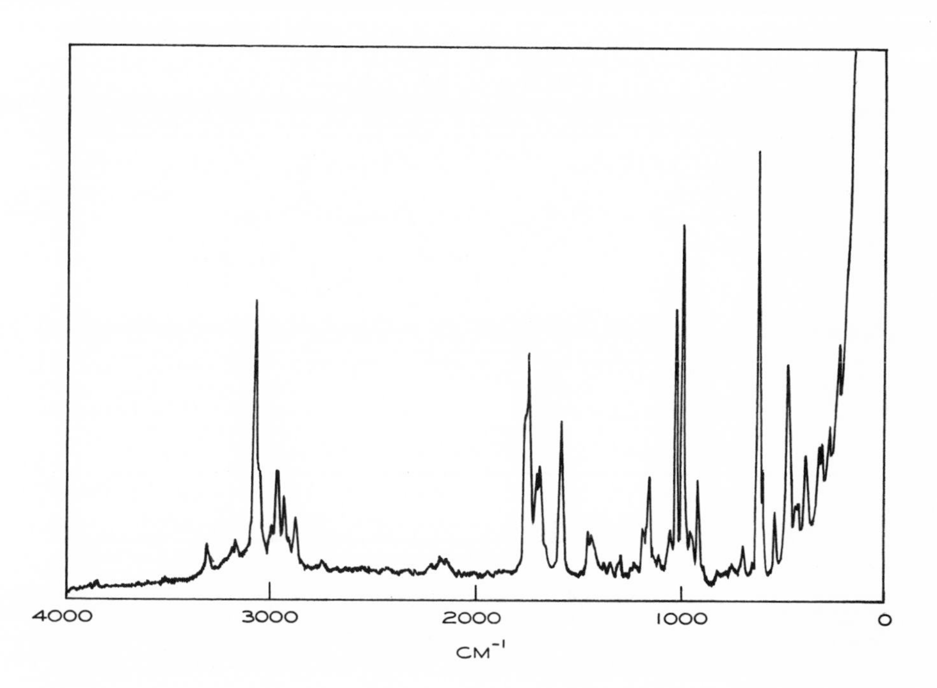

Figure 6.3. Raman spectrum of phenobarbital (7).

Musks, important perfume constituents and possibly of biological importance in man (14), have some relatively simple structural criteria in common, that is, a molecular weight between about 210 and 290, a bulky, closely packed and rigid profile, and a sterically accessible functional group. The fact that the isochroman derivative I in Figure 6.4 is a strong musk, while the 1- and 3-methyl isomers (II and III) are odorless, has been attributed to a steric effect operating in the latter two compounds that severely reduces the accessibility of the oxygen atom (15).

The Raman spectra of I and V differ from II-IV by the appearance of a band in the spectrum of the liquid which is absent in the spectrum of the solid (Figure 6.5). In contrast, the spectra of II and III, as well as the apomethyl compound IV, show no differences in this region between the liquid and solid states. Depolarization values of approximately 0.03 were obtained on these bands for all four materials in the liquid state.

TABLE 6.12. Characteristic Raman Frequencies of the Hypoxanthine Ring System (8)

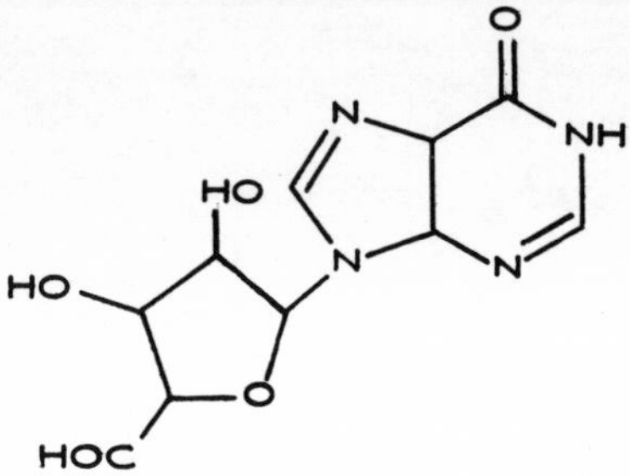

Solution	Inosine	1-Methyl-Inosine-5'-P	1-Methyl Inosine	Rel. Int.
Acidic	720±3	720±2	718±3	m
	1034±2	1034±2	1040±2	m
	1386±3	1386±2	1382±5	m
	1425±6	1423±5	1437±7	s
	1533±4	1531±2	1536±3	s
	1570±6	1570±6	1573±3	vs
Neutral	722±2	723±2	721±2	s
	1046±2	1053±2	1048±2	m
	1336±2	1336±2	1335±2	vs
	1386±4	1386±4	1388±2	m
	1423±2	1424±4	1434±2	m
	1552±3	1551±3	1555±2	vs
	1587±7	1588±6	1586±2	m
Alkaline	738±2	738±2		s
	1051±2	1051±2		m
	1140±2	1139±2		m
	1302±2	1301±2		s
	1336±2	1336±2		vs
	1376±2	1376±2		m
	1420±2	1420±2		m
	1475±2	1474±2		vs
	1564±2	1562±2		vs
	1594±2	1593±2		m

This general spectral region for isochroman, 1-methyl isochroman, and 3-methyl isochroman (Figure 6.6), all as liquids, shows a single band for each compound ($\rho \sim 0.03$).

Liquid-- 682, 703 cm^{-1}
Solid-- 682 cm^{-1}

I

Liquid and Solid-- 702 cm^{-1}

II

Liquid and Solid-- 685 cm^{-1}

III

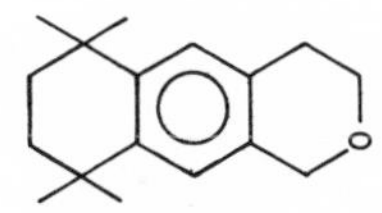

Liquid and Solid-- 683 cm^{-1}

IV

Liquid-- 710, 730 cm^{-1}
Solid-- 710 cm^{-1}

V

I, V -- Odorous

II, III, IV -- Odorless

Figure 6.4. Raman spectral data for odorous and odorless isochromans; I,V-odorous; II,III,IV-odorless (13).

On the other hand, two bands are evident for 4-methyl isochroman (ρ = 0.02). The frequencies, relatively high scattering efficiences, and low depolarization values of the nine isochromans pictured in Figures 6.4 and 6.6 are consonant with a ring breathing mode of the dihydropyran moiety.

A study of Dreiding Stereomodels (Figure 6.7) shows that the odorous musk I as well as V (Figure 6.4) can adopt two conformations in which the orientation of the 4-methyl group is *cis* and *trans* with respect to the C-3—O bond (1). Neither of these is sterically inhibited by 1,3-interaction. Two conformations each are also possible in II and III, but here the conformations with the 1- and 3-methyl groups in the quasi-axial position are hindered by 1,3-interaction (II "*trans*", Figure 6.7). Such is not the case for the conformers in which the 1- and 3-methyl groups assume the quasi-equatorial positions (II "*cis*"), and the Raman spectral data suggest that the 1- and 3-methyl isomers exist in this conformation only. The steric accessibility of the oxygen atom is reduced considerably in both the 1- and 3-methyl compounds, correlatable with the fact that they are odorless.

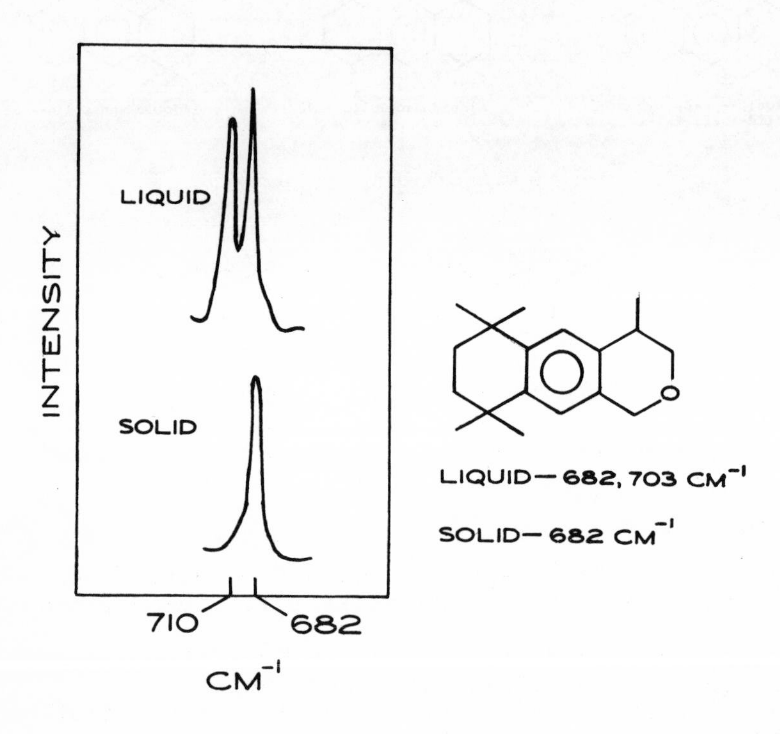

Figure 6.5. Indication of conformeric states by Raman spectroscopy (13).

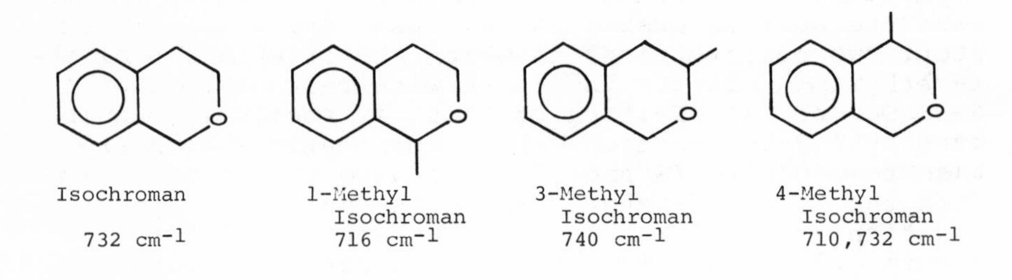

Figure 6.6. Strong, highly polarized ($\rho \sim 0.03$) bands for some simple isochromans (13).

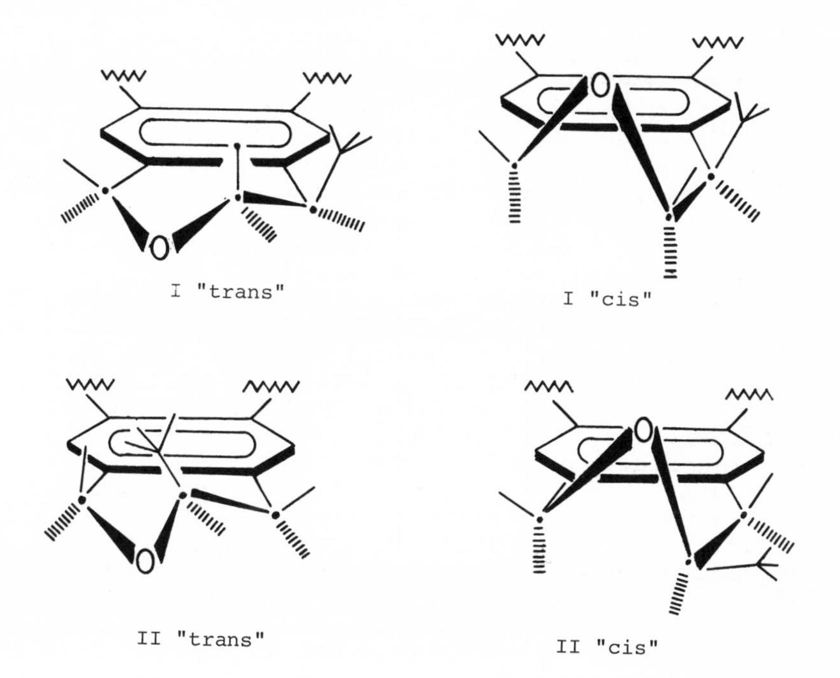

Figure 6.7. Steric considerations in odorous and odorless isochromans.

However, spectral evidence suggests that two conformers are present in the 4-methyl compounds (I and V, Figure 6.4). According to the molecular models, the steric accessibility of the oxygen atom is much greater in I "*trans*" than in I "*cis*" (Figure 6.7), and it may be assumed that the intense musk odor of the 4-methyl derivatives is attributable mainly, if not totally, to the presence of I "*trans*."

6.1.7 Steroids

Prior to the middle of this century the numerous structural investigations of steroids were limited essentially to the laborious classical methods of chemical degradation followed by synthesis. After the steroid framework became known in 1932, the structures of many steroids were elucidated in rapid succession up to 1937 and stereochemical studies led to assignment of absolute configurations in 1955. Research in this area gained impetus from the discovery that many steroids were therapeutic agents. In addition to the naturally occuring materials, many new compounds were synthesized. By this time, it was no longer sufficient to apply chemical methods alone to determine steroid structures; molecular spectroscopy, for example, X-ray, UV, IR, MS, NMR, ORD, and CD, was successfully employed to achieve this goal. However, except for X-ray diffraction, these methods yielded little information regarding the steroid skeleton. Recently, it was demonstrated that Raman spectroscopy is an excellent probe for the molecular framework as well as for most functional group characterizations (16). The basic skeletons of naturally occurring steroids differ only in the type of fusion of rings A-B and C-D (Figure 6.8). Rings C and D are *trans* except for the heart glycosides and toad poisons. The Raman spectra of steroids are complex (Figure 6.9). A procedure has been developed for the structural elucidation of steroids (16), based on a computerized analysis of 80 crystalline steroids belonging to the androstan, pregnan, and estran series. This approach takes into consideration positions and relative intensities of Raman bands (Tables 6.13 and 6.14). Caution should be exercised in using only these data in the case of an "unknown" material. The intensity terms strong, normal, and so on, refer to the strength of a particular band (I) relative to the intensity at 1450 cm^{-1} (I_{1450}). Very strong, or very intense, means $I = 5 \times I_{1450}$; strong signifies $I = 2 \times I_{1450}$; normal means $I = I_{1450}$; medium signifies $I = (1/2) \times I_{1450}$; and weak means $I = (1/5) \times I_{1450}$.

In addition to valuable information gained from frequencies and intensites of C=O and C=C stretching vibrations (1600-1725 cm^{-1}), the region between 800 and 1350 cm^{-1} offers important interpretive clues. Intense IR absorptions arise from C-O vibrations of hydroxyl, ester, and ether groups, but no IR bands are found which are associated with particular carbon-carbon double bonds of 3-keto steroids. Strong Raman bands near 960 cm^{-1}, as well as scattering at about 1240 and 1280 cm^{-1}, are diagnostic for 1- and 4-unsaturated compounds.

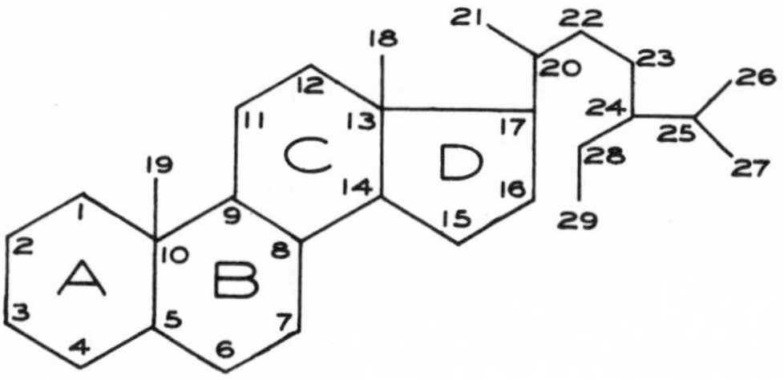

Basic skeleton

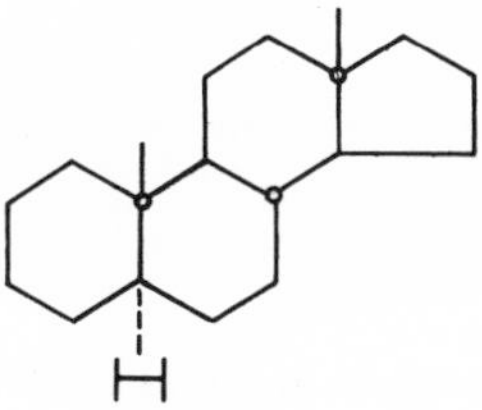

5α Steroid
(A/B trans)

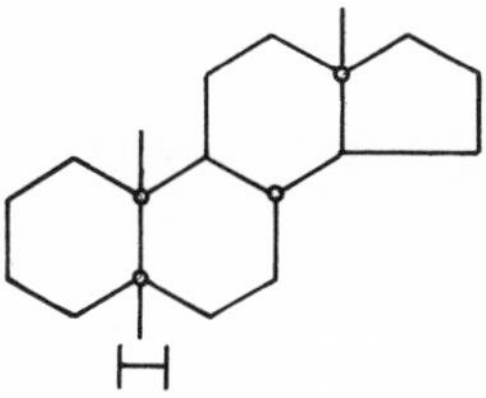

5β Steroid
(A/B cis)

Figure 6.8. Steroid structures.

The 1,4-dienic-3-keto steroids are recognized by strong bands near 1110 and 1165 cm^{-1}, while 4,6-dienic analogs have intense bands at 1200 and 1230 cm^{-1}. These data support conclusions arrived at from examination of the C=C stretching region (see p.111).

A band of moderate intensity is observed at about 500 cm^{-1} in the spectrum of 5β-steroids, but is absent in the case of 5α-steroids. Bands of varying intensity appear between 40 and 150 cm^{-1} in the Raman spectra of crystalline steroids. The intensity of these lattice vibrations increases in the order: saturated < unsaturated < conjugated keto < aromatic ring-containing steroids.

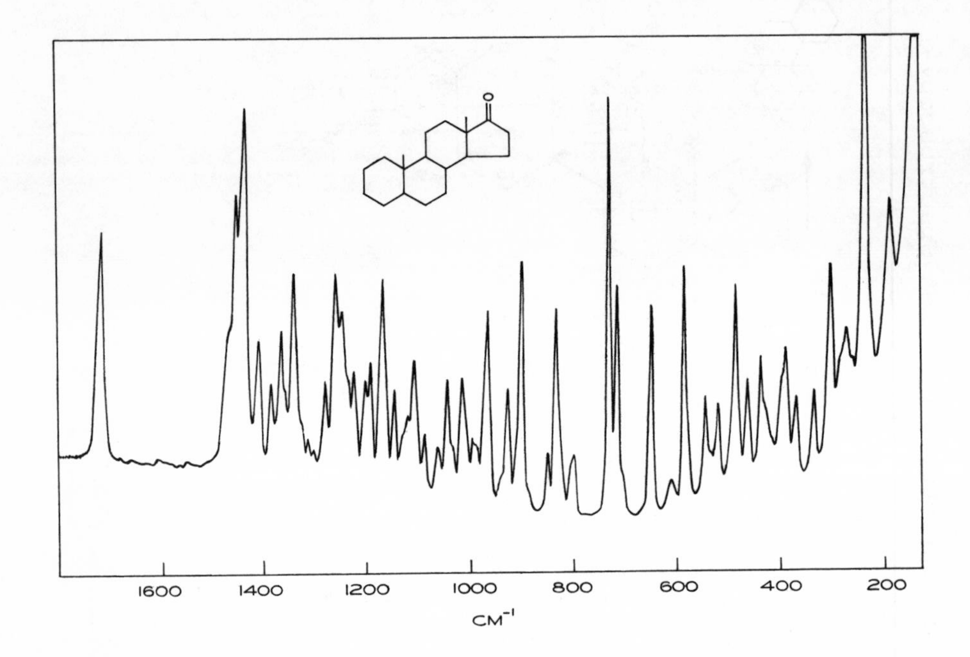

Figure 6.9. Raman spectrum of androsterone.

6.1.8 General Considerations

Besides yielding information pertaining to the ring system type of an "unknown" material, Raman spectroscopy often is helpful in monitoring reactions which involve cyclization, rearrangement, oxidation, and so forth.

Cyclization

Cyclic compounds often are formed from acyclic intermediates. The appearance of a strong band or a group of moderate to intense bands between about 400-1000 cm^{-1} may be used to track such cyclizations. For example, treatment of nerol with an organic acid produces alpha terpineol (1) (Figure 6.10), accompanied by the appearance of an intense 756 cm^{-1} band (ρ = 0.02).

TABLE 6.13. Procedure to Determine Steroid Structures by Raman Spectroscopy-I (16)

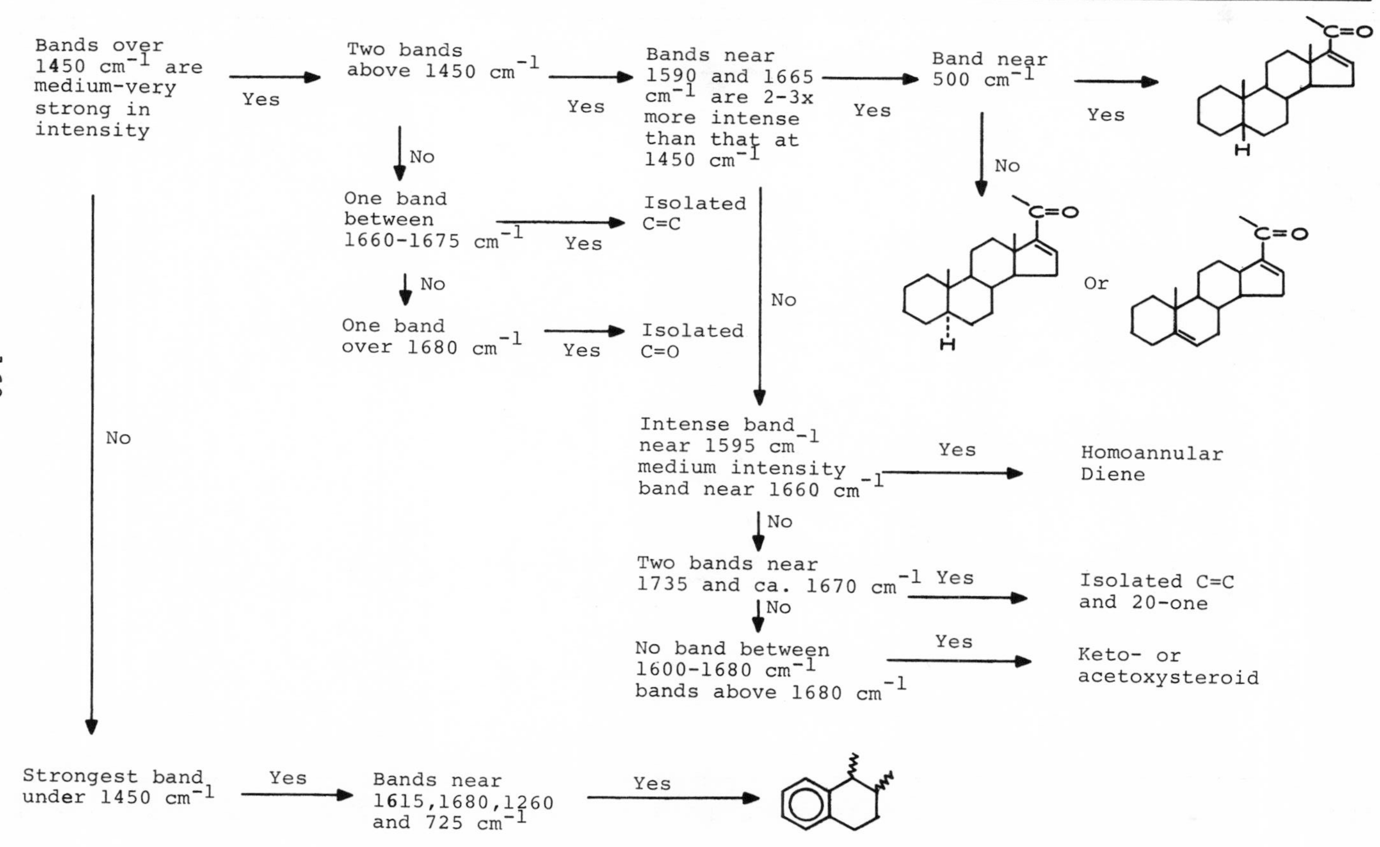

TABLE 6.14. Procedure to Determine Steroid Structures by Raman Spectroscopy-II (16)

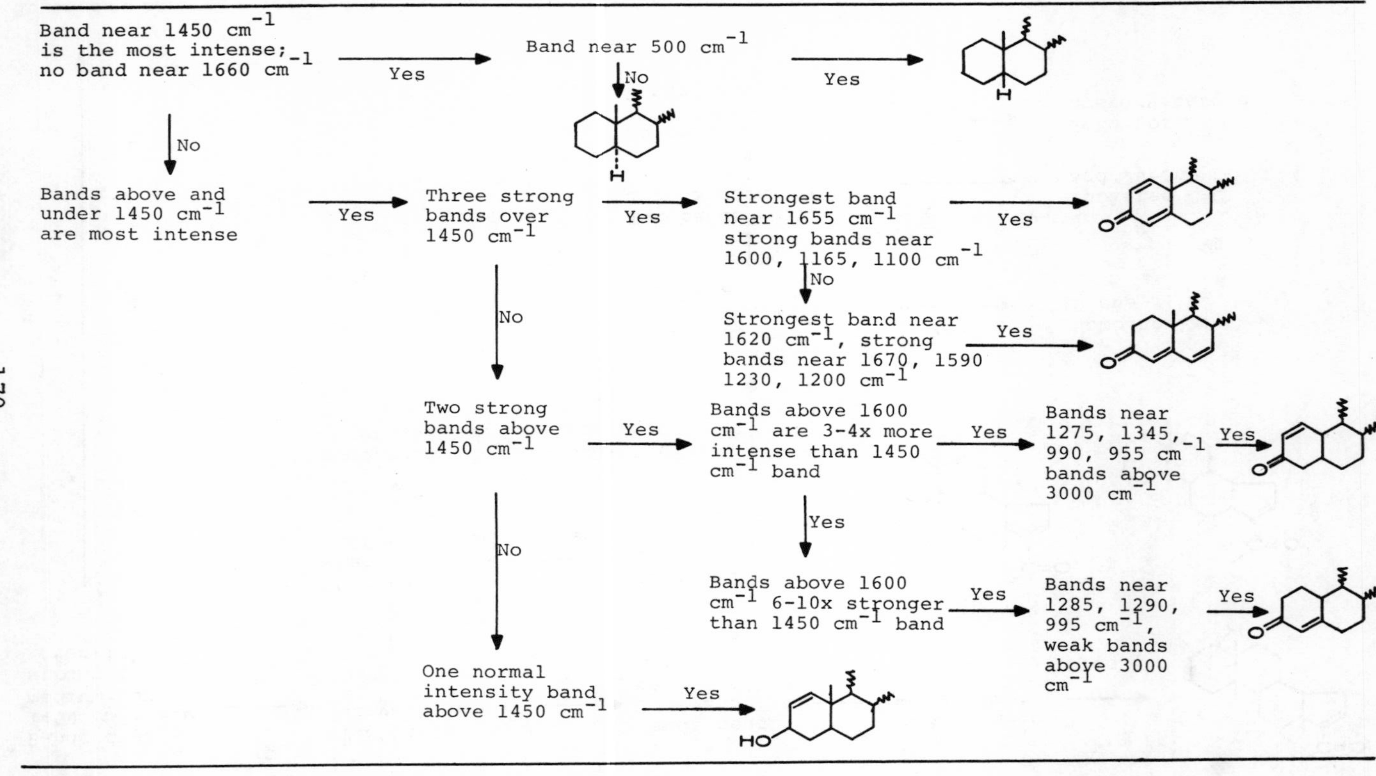

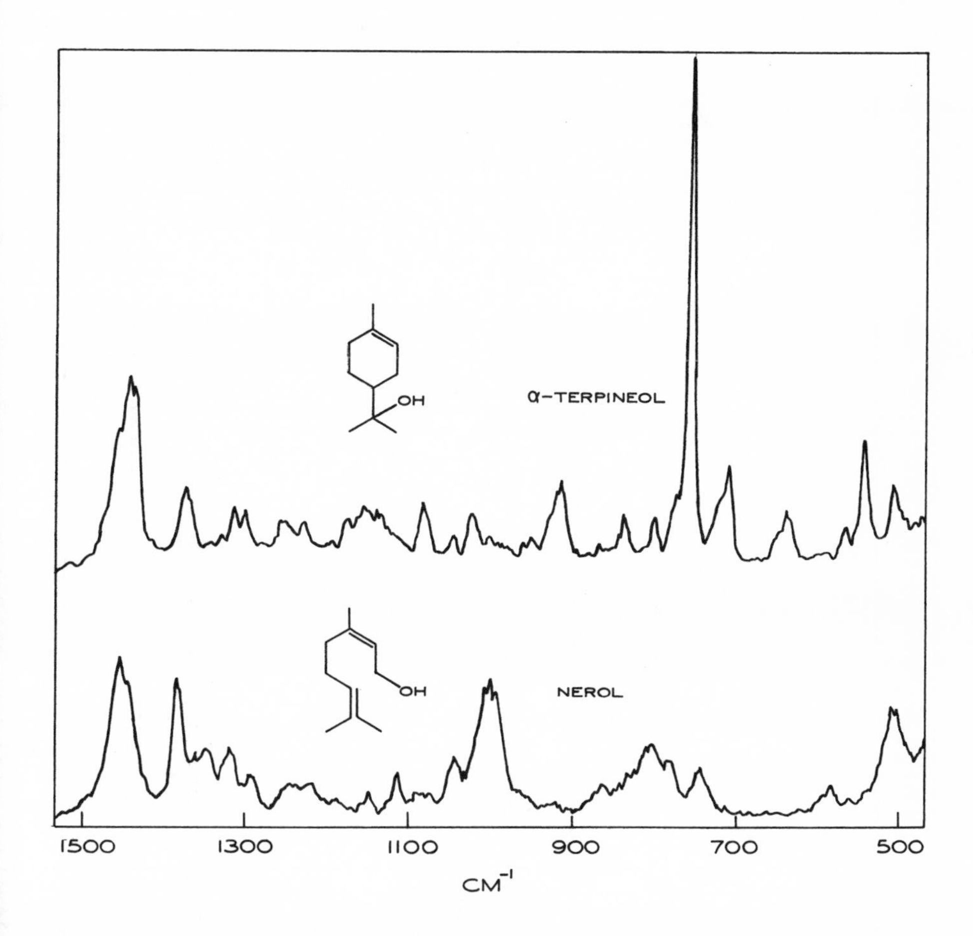

Figure 6.10. Raman spectra of nerol and its cyclized product, α-terpineol. Note the strong ring breathing band at 756 cm^{-1} (ρ = 0.02).

Ring System Rearrangement

Treatment of a bi- or tricyclic compound with particular reagents sometimes brings about a skeletal rearrangement. Under the influence of an organic acid, longifolene is transformed into isolongifolene (1) (Figure 6.11).

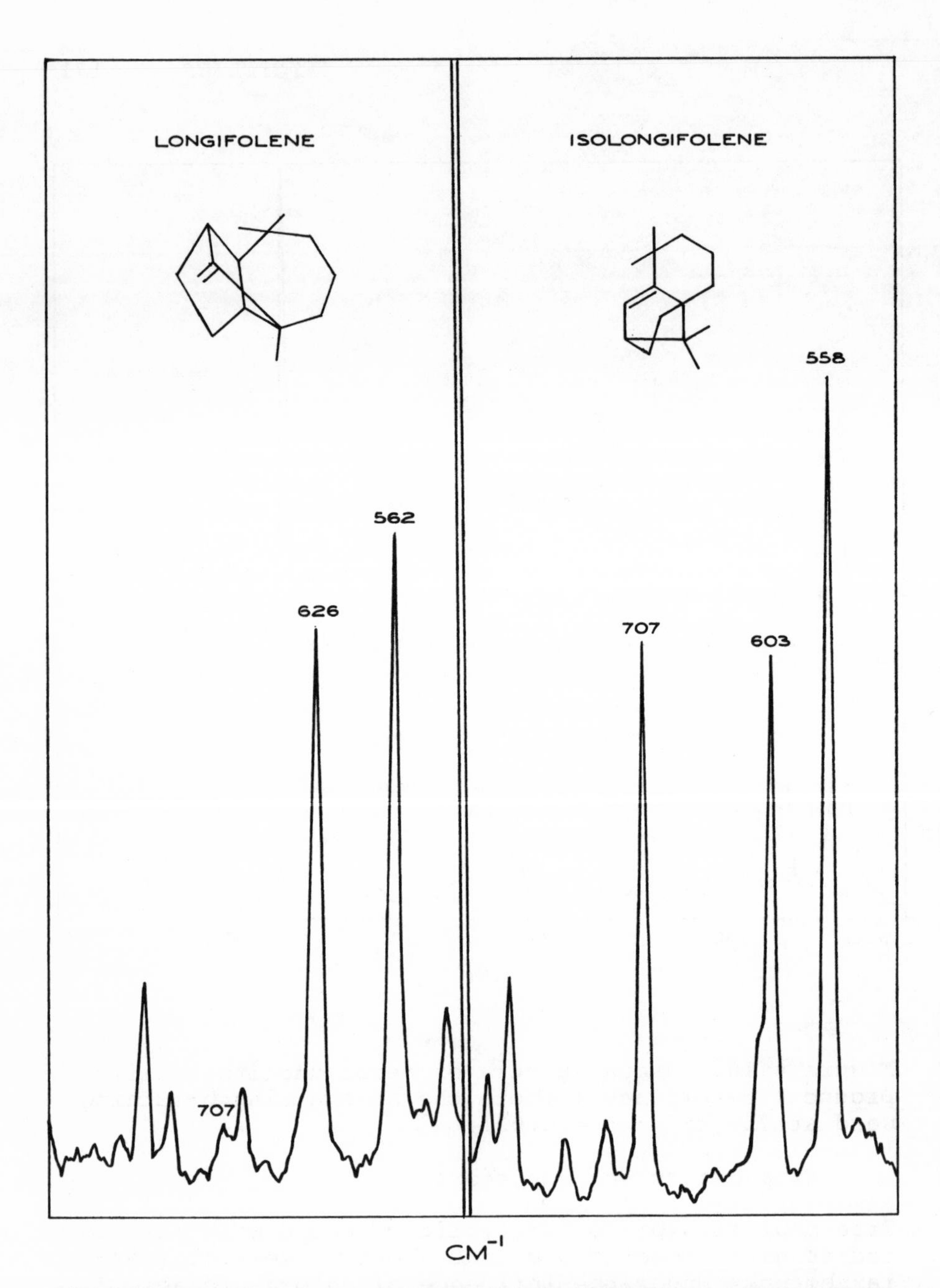

Figure 6.11. Raman spectra of longifolene and isolongifolene. Note absence of 707 cm^{-1} (ρ = 0.02) band in former.

Although it might be argued that information regarding the rearrangement could be obtained just as easily from the C=C stretching region ($\nu_{C=CH_2}$ = 1650 cm^{-1}, $\nu_{C=CH}$ = 1675 cm^{-1}), the isolongifolene band at 707 cm^{-1} is approximately three times more intense than the C=C stretching bands, making the former a more sensitive probe.

Another application of Raman spectroscopy in ring rearrangements is found in patchouli compounds (1). Pyrolysis of patchouli acetate at 300°, or dehydration of the alcohol, affords a mixture of α, β, and γ-patchoulenes (Figure 6.12).

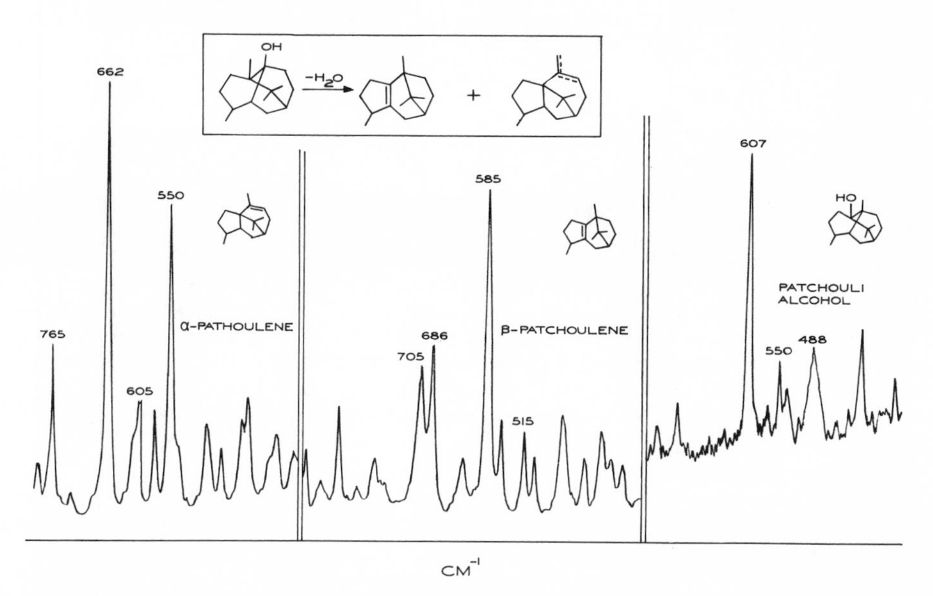

Figure 6.12. Raman spectra of patchouli alcohol and sesquiterpenes formed on dehydration. γ-Patchoulene resembles the α-isomer between 500 and 700 cm^{-1}.

Raman spectra of the gas chromatographic isolates indicate that the three different ring systems can be spectrally distinguished: α- and γ-patchoulenes are similar between 500 and 700 cm^{-1}, and they are different from β-patchoulene and patchouli alcohol. Examination of the 500-800 cm^{-1} region for a few derivatives of each ring system suggests that the three patchouli ring types have specific bands which allow their general differentiation (1).

Caution must be exercised in adopting this approach in the absence of data on several compounds possessing the same ring system because obvious spectral features may be similar for different ring systems. A case in point is adamantane. The most intense band of adamantanone (VI) lies between 700 and 800 cm^{-1}, but protoadamantane (VII) also displays its strongest band in this region (1).

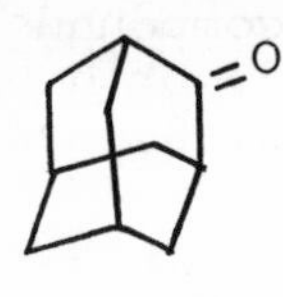

VI

Adamantanone

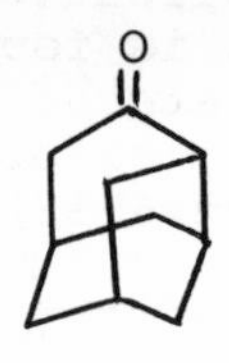

VII

Protoadamantanone

Ring Substituents

Since the nature of a ring substituent usually has little effect on the ring breathing frequency(s) appearing between 500 and 1000 cm^{-1}, the band(s) can be used as a diagnostic for ring integrity during reactions involving substituents. For example, oxidation of ring tertiary alcohols to ketones with chromic acid may proceed to the acids. The strong ring vibration band(s) decrease in intensity with increasing keto acid concentration.

6.2 AROMATICS

The characterization of aromatic rings has been a classic application of infrared spectroscopy for more than two decades. Consequently, relatively little attention has been given to the potential role of Raman spectroscopy in this area, in spite of the fact that the groundwork had been laid during the 1930s (17).

Most aromatics show one or more moderate to strong Raman bands near 1600 cm^{-1} due to ring deformation modes that correspond to quadrants of the ring stretching and contracting (18). They are often observed in IR spectra, but with low intensity. Other Raman and

IR bands must be employed for determining the location of a substituent(s) on an aromatic ring. The Raman spectra of aromatics usually are simpler, that is, they contain fewer bands than their IR counterparts, and this fact eases the task of interpreting Raman data. Since the type of aromaticity, that is, benzenic, pyridinic, pyrazinic, thiophenic, furanic, naphthalenic, and so on, at times cannot be gleaned readily from the vibrational spectra of a compound whose history is unknown, it is necessary to consider other spectral data. Ultraviolet absorption curves are quite helpful in this respect (Figure 6.13). Conjugation of a ketone or aldehyde with aromatic ring systems causes changes in UV profiles. Obviously, the empirical formula and degree of unsaturation derived from the mass spectrum are additional clues to aromatic nature, and proton NMR chemical shifts aid determining aromatic type and substituent location.

6.2.1 Benzenes

Monosubstituted

For compounds of this class, the IR out-of-plane hydrogen deformation and ring puckering modes near 750 and 700 cm^{-1}, respectively, are excellent diagnostics for a monosubstituted benzene ring (Figure 6.14) (18). Nevertheless, there are several substituents for which this correlation does not hold and may cause confusion in interpretation. Some of these are benzyl halides, benzoate esters, 4-methyl-2-phenyl pentanal, and nitrobenzene (Figure 6.15). On the other hand, three characteristic lines appear in the Raman spectra of these and all other monosubstituted benzenes examined: a very strong symmetric ring stretch at 997 ± 8 cm^{-1}, a moderate strength in-plane CH bending band at 1028 ± 7 cm^{-1} and a weak, depolarized in-plane ring deformation band at 615 ± 10 cm^{-1} (Figure 6.16) (18).

Disubstituted

Table 6.15 lists the diagnostically important bands for ortho, meta, and para disubstituted benzenes. The band pattern for ortho derivatives resembles the monosubstituted benzenes, but differs in that only one strong, polarized Raman band appears between 1015-1050 cm^{-1}

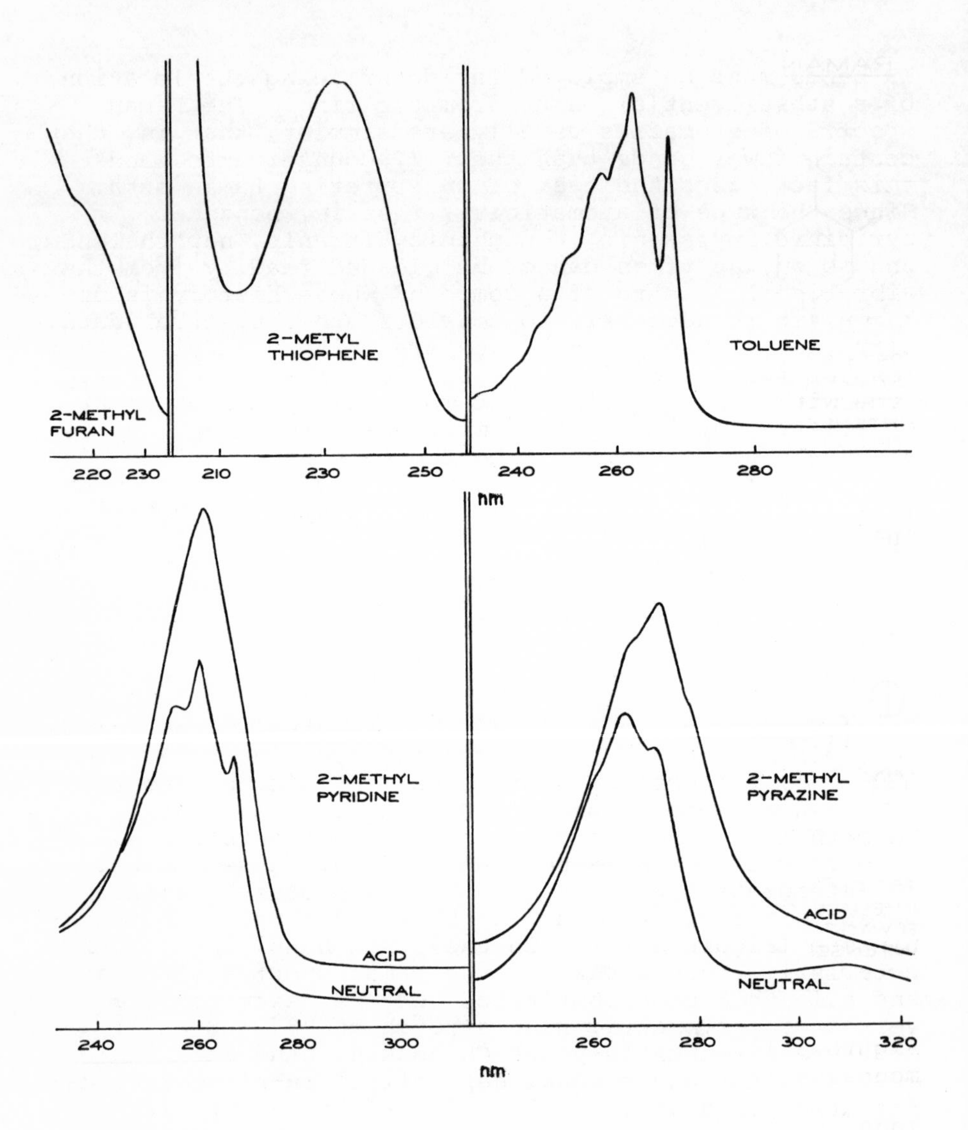

Figure 6.13. Ultraviolet absorption spectra of various aromatic systems.

instead of a strong to medium pair. Furthermore, the depolarized band observed in the 700 cm^{-1} region is at higher frequency (700 ± 60 cm^{-1}) than in monosubstituted benzenes (615 ± 10 cm^{-1}). The absence of a band near

RAMAN

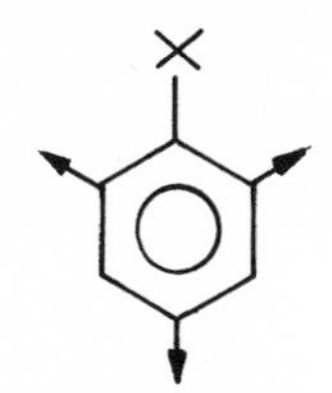

997 ± 8 CM^{-1}
SYMMETRIC RING STRETCH
INTENSE
(POLARIZED)

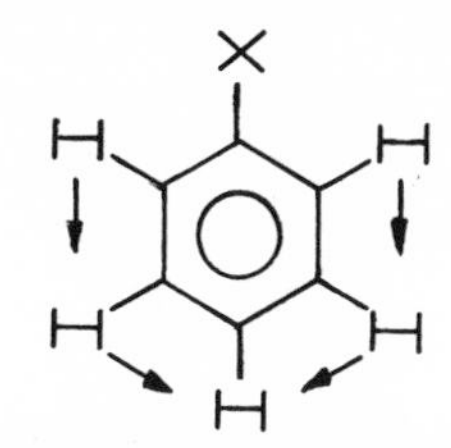

1022 ± 8 CM^{-1}
IN-PLANE H BEND
MODERATE

615 ± 10 CM^{-1}
IN-PLANE RING DEFORMATION
WEAK
(DEPOLARIZED)

IR

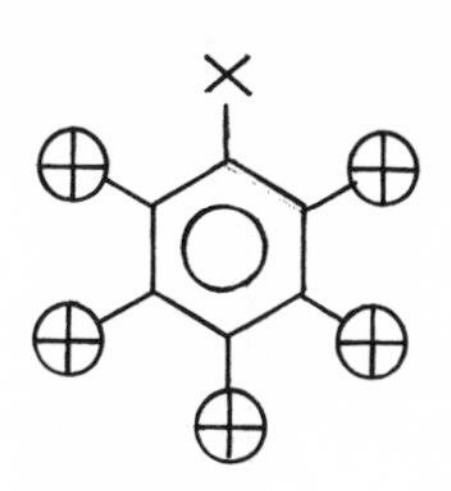

750 CM^{-1}
IN-PHASE, OUT-OF-PLANE H WAG
INTENSE

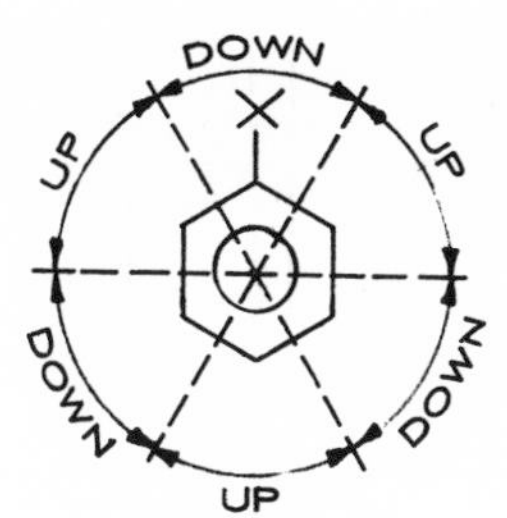

700 CM^{-1}
RING PUCKER
INTENSE

Figure 6.14. Characteristic Raman and IR bands for monosubstituted benzenes (18).

1000 cm^{-1} is characteristic of para disubstituted benzenes, and a weak band is generally observed between 625 and 645 cm^{-1}. Similar to the ortho derivatives, meta disubstituted benzenes display one band only near 1000 cm^{-1}, but at lower frequencies (990-1010 cm^{-1} compared to 1020-1050 cm^{-1}). Also, no bands for the meta compounds occur in the 650-750 cm^{-1} region. When meta directing groups are substituted on a benzene ring, the

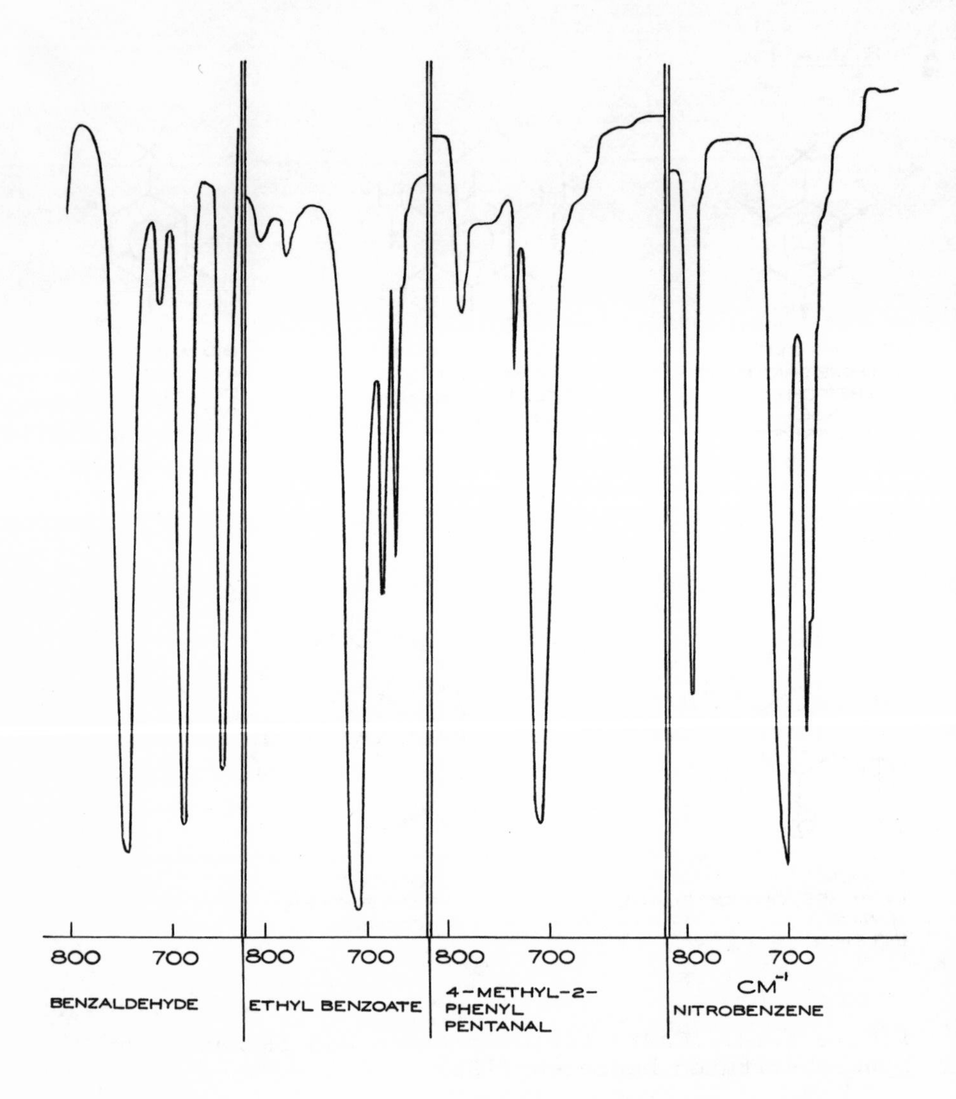

Figure 6.15. Some monosubstituted benzenes which do not display the diagnostic IR hydrogen deformation ($\sim$750 cm^{-1}) and ring puckering ($\sim$700 cm^{-1}) bands (benzaldehyde behaves in a normal fashion).

out-of-plane hydrogen vibrations which give rise to characteristic IR bands are greatly perturbed (19).

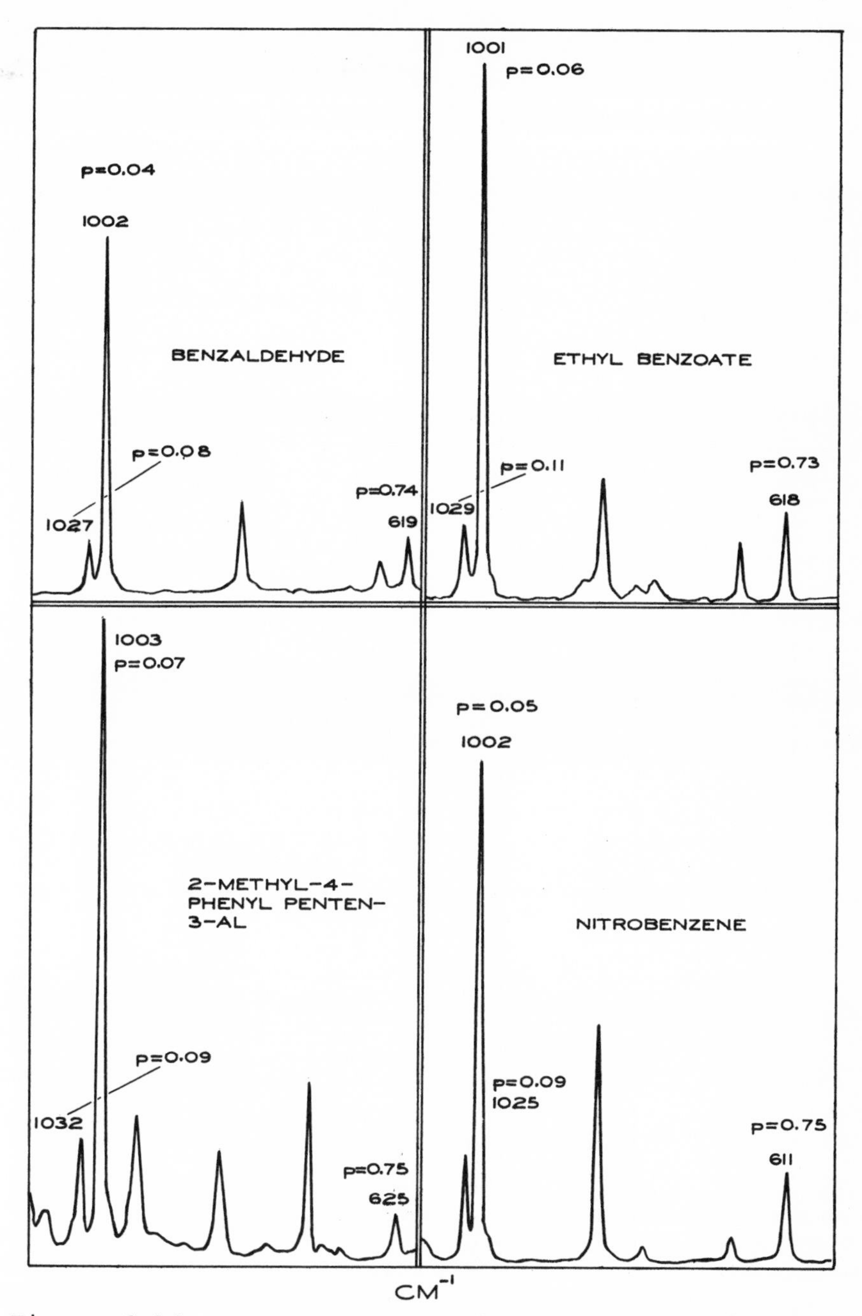

Figure 6.16. Raman spectra of some monosubstituted benzenes exhibiting anomalous IR spectra between 650 and 800 cm^{-1}.

TABLE 6.15. Characteristic Raman Bands for Simple *Ortho*-, *Meta*-, and *Para*-Disubstituted Benzenes

Ortho (10 Compounds)			Meta (8 Compounds)			Para (12 Compounds)		
Raman (cm^{-1})	ρ	IR[a]	Raman	ρ	IR[a]	Raman	ρ	IR
1035±15[b]	∿0.03	750±8	1000±10[c]	∿0.1	782±9	635±10[d]	∿0.2	817±13
700±60					690±15			

[a]Out-of-plane H wag.

[b] (X, Y, H, H, H, H substituted benzene ring) In-plane H deformation (same ν for symmetric trisubstituted benzene derivatives).

[c]Related to 997±8 cm^{-1} ring stretch in monosubstituted benzenes.

[d]Related to 615±10 cm^{-1} in-plane ring deformation in monosubstituted benzenes.

In many instances, the intensities of bands associated with these vibrations decrease and the bands also are shifted to higher frequencies into regions where other vibrations occur. Consequently, it is extremely difficult to determine ring substitution of this type by IR. No such anomalies are encountered in the Raman effect.

6.2.2 Pyridines

IR spectroscopy is acknowledged to be a powerful tool for characterizing pyridines (20,21) and recently the potential of the Raman effect has been recognized for this class of compounds (1). Resembling benzene derivatives (see p.175), ring breathing, in-plane hydrogen deformation, and substituent sensitive vibrations in pyridines give rise to interpretively significant Raman bands. (See Table 6.16.)

TABLE 6.16. Characteristic Raman Bands for Monosubstituted Pyridines ($\rho \sim 0.04$)

2-Substituted (10 Compounds)		3-Substituted (8 Compounds)		4-Substituted (11 Compounds)	
cm^{-1}	Int.	cm^{-1}	Int.	cm^{-1}	Int.
825±25[a]	1-6	777±27[a]	1-5	795±6[a]	2-8
998±2 [b]	9-10	1040±10[c]	10	997±2[b]	10
1049±2 [c]	6-10				

[a]Substituent sensitive.

[b]Ring breathing.

[c]In-plane hydrogen deformation.

6.2.3 Pyrazines

Recent findings that pyrazines are important flavor constituents of many natural and heat-processed foods prompted Raman investigations of a host of pyrazines (22-24). Raman spectra-structure correlations now allow simple and unequivocal determination of many substitution patterns and, in some instances, components of a pyrazine mixture. Vibrational modes and characteristic data for substituted pyrazines are shown in Tables 6.17-6.20 (23) and Table 6.21.

6.2.4 2-Substituted Furans

Pre-laser Raman studies on some 2-substituted furans described several bands characteristic of the aromatic nucleus. The data have been confirmed recently (Table 6.21), and band frequencies were found to differ significantly from those observed for 3-, 5-monosubstituted and 2,5-disubstituted derivatives (1).

6.2.5 2-Substituted Thiophenes

A band pattern for 2-substituted thiophenes, differing from those present in the Raman spectra of 3-, 5-monosubstituted and 2,5-disubstituted compounds, suggests that it may be of interpretive value for this class of thiophenes (1,28) (Table 6.21) (1).

6.2.6 2-Pyrroles

Raman spectra of several 2-substituted pyrroles show two bands in common that are not observed in the spectra of a few N-, 3-, 5-monosubstituted and 2,5-disubstituted pyrroles (Table 6.22) (1).

6.2.7 Naphthalenes

Analogous to other aromatic rings, IR correlations are known for the in-phase, out-of-plane hydrogen wagging vibrations in substituted naphthalenes (29,30). Raman spectroscopy has been demonstrated to be a valuable probe for 1- (References 1,30, and 31) and 2- (References 1,31, and 32) substituted naphthalenes (Table 6.22) (1). A characteristic, strong band appears between 1370 and 1390 cm^{-1} ($\rho \sim 0.1$), being present in Raman spectra of 1- and 2-monosubstituted, 1,2-, 1,3-, 1,4-, 1,5-, 1,8-, and 2,6-disubstituted compounds (1,31,33).

TABLE 6.17. Vibrational Modes for Monosubstituted Pyrazines (20 Compounds) (23)

cm^{-1} Approx.	ρ Approx.	Intensity	Vibrational Mode[a]
640	0.5	Weak	(I)
790	0.04	Strong	(II)
1015	0.04	Very strong	(III)
1055	0.06	Strong	(IV)
1575	0.20	Medium	(V)

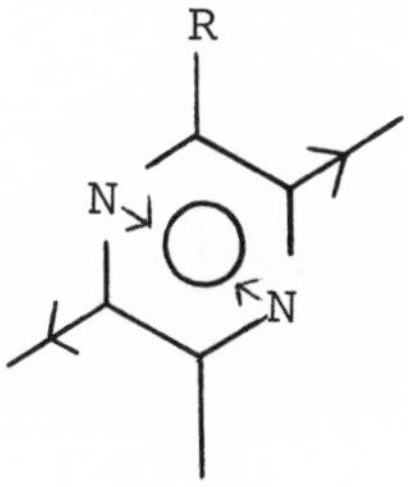

R N N

I

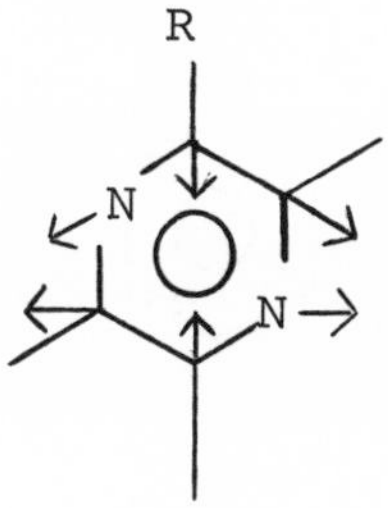

R N N

II

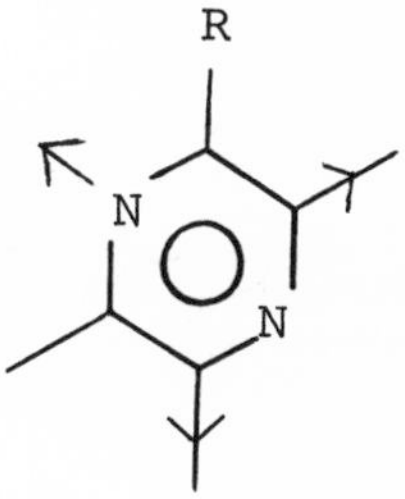

R N N

III

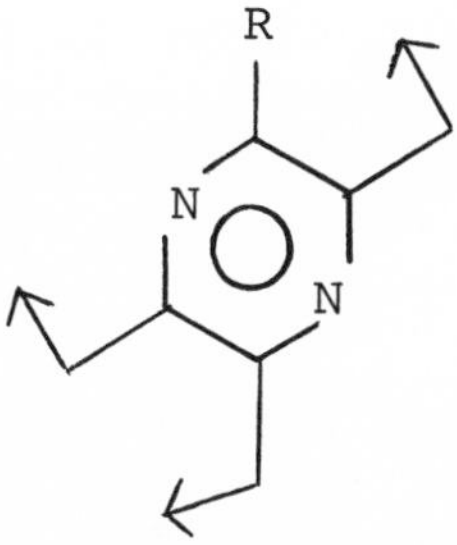

R N N

IV

R N N

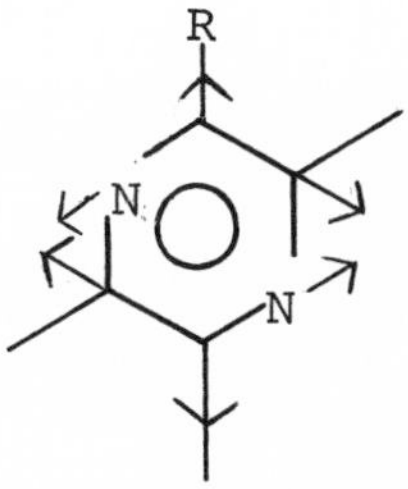

V

[a]Analogous vibrations (cm^{-1}):
Toluene: 623 (I), 786 (II), 1044 (III), 1041 (IV), 1604 (V)
2-Picoline: --, --, 994 (III), 1047 (IV), 1590 (V)

TABLE 6.18. Vibrational Modes for 2,3-Disubstituted Pyrazines (15 Compounds) (23)

cm^{-1} Approx.	ρ Approx.	Intensity	Vibrational Mode[a]
720	0.05	Very strong	(I)
1085	0.1	Very strong	(II)
1275	0.18	Medium	(III)

I II III

[a]Analogous vibrations (cm^{-1}):
o-Xylene: 733 (I) 1052 (II) 1275 (III)
2,3-Lutidine: 730 (I) 1073 (II) 1292 (III)

6.2.8 The 1,2-Dithiole Ring

The physical and chemical properties of 1,2-dithiole-3-thione (VIII) and 1,2-dithiole-3-one (IX) indicate that they are pseudoaromatic molecules (34).

TABLE 6.19. Vibrational Modes for 2,5-Disubstituted Pyrazines (12 Compounds) (23)

cm^{-1} Approx.	ρ Approx.	Intensity	Vibrational Mode[a]
647	0.5	Medium	(I)
845	0.04	Very strong	(II)
1220	0.05	Strong	(III)

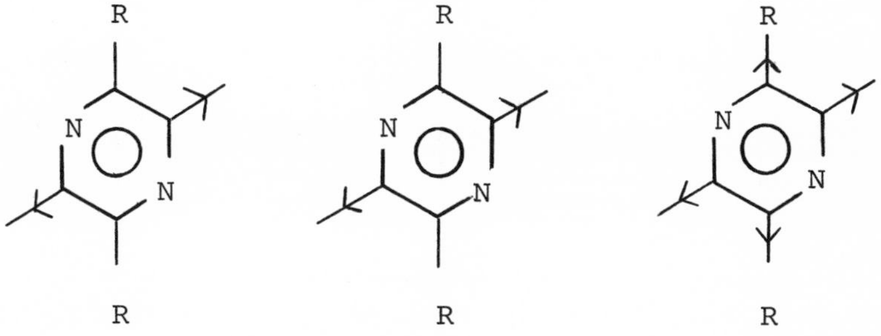

I II III

[a]Analogous vibrations (cm^{-1}):
p-Xylene: 643 (I) 828 (II) 1182 (III)
2,5-Lutidine: 646 (I) 840 (II) 1220 (III)

Although the S-S bond distance (2.04 Å) has approximately the value found for two sulfur atoms in an alkyl disulfide (2.04 Å) these structures are quite different. In the 1,2-dithiole ring, the C-S-S-C atoms are coplanar, as compared with a dihedral angle of about 90° for alkyl disulfides (see p.218). Characteristic Raman frequencies for the dithiolic ring, based on five compounds, are listed in Table 6.22 (35).

TABLE 6.20. Vibrational Modes for 2,6-Disubstituted Pyrazines (5 Compounds) (23)

cm^{-1} Approx.	ρ Approx.	Intensity	Vibrational Mode[a]
708	0.03	Very strong	(I)
1023	0.05	Strong	(II)
1163	0.02	Medium	(III)

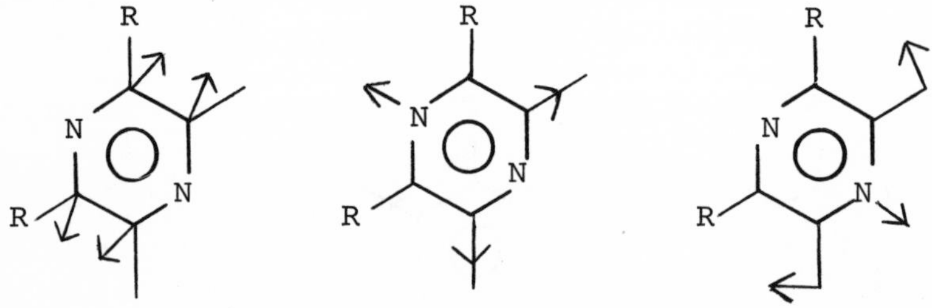

R R R
N N N
N N N
R R R

[a]Analogous vibrations (cm^{-1}):
m-Xylene: 724 (I) 999 (II) 1094 (III)
2,6-Lutidine: 720 (I) 997 (II) 1155 (III)

6.2.9 Purine and Pyrimidine Bases

These are shown in Table 10.8, pg.289.

TABLE 6.21. Characteristic Raman Data for Pyrazines (22,23)

	ν (cm^{-1})	ρ	Intensity
Monosubstituted	617-660	0.3-0.6	1-3
	788-840	0.02-0.06	2-10
	1003-1024	0.02-0.08	8-10
	1050-1060	0.02-0.1	2-10
	1517-1530	0.4-0.7	1-5
	1560-1580	0.14-0.24	1-4

TABLE 6.21. (Continued)

	ν (cm^{-1})	ρ	Intensity
2,3-Disubstituted	686-758	0.03-0.07	10
	1081-1100	0.1-0.2	1-8
	1252-1292	0.15-0.2	1-10
	1528-1570	0.5-0.7	1-7
	1558-1580	0.1-0.2	1-7
2,6-Disubstituted	706-709	0.03	10
	1021-1025	0.04	6-7
	1572-1586	--	1-3
2,5-Disubstituted	642-650	0.5	2-3
	838-865	0.04	10
	1520-1540	0.6	1-7
	1576-1586	0.2	2-4
Trisubstituted	695-710	--	8-10
	725-748	0.05	2-4
	915-955	--	2-4
	1525-1540	0.6	3-6
	1570-1576	0.3	3-10
Tetrasubstituted	710-720	--	7-10
	1545-1550	--	3-10
	1560-1565	--	1-3

TABLE 6.22. Characteristic Raman Data for Various Aromatics

Compound	ν (cm^{-1})	Approx	Rel. Int.
2-Substituted furans (14 compounds)	1065-1090	0.1	2-6
	1382-1395	0.05	1-4
	1475-1515	0.1	10
	1565-1605	0.2	2-3
2-Substituted thiophenes (10 compounds)	632-680	0.03	1-8
	845-860	0.1	1-5
	1080-1040	0.15	1-9
	1405-1440	0.25	10

TABLE 6.22. (Continued)

Compound	ν (cm^{-1})	Approx.	Rel. Int.
1-Substituted naphthalenes (10 compounds)	650-720	0.05	2-10
	810-880	0.08	1-5
	1020-1080	0.1	1-3
	1370-1385	0.3	10
2-Substituted naphthalenes (8 compounds)	515-525	0.02	2-6
	765-775	0.03	8-10
	1380-1390	0.1	10

TABLE 6.23. Characteristic Raman Data for the Dithiolic Ring

ν (cm^{-1})	ρ (Approximate)	Intensity	Assignment
1495-1546	0.2	m-vs	Ring vibration
1264-1330	0.2	m-vs	Ring vibration
1171-1180	0.3	m-vs	C=S stretch
982-1044	0.3	m-s	Ring vibration
472-585	0.2	s-vs	Ring vibration

REFERENCES

1. S. K. Freeman (unpublished work).
2. Am. Petr. Inst. Proj. 44, Chem. Thermo. Properties Center, A & M College of Texas, College Station, Texas.
3. P. A. Bazhulin, H. E. Sterin, T. F. Bulanova, O. P. Solovova, M. B. Turova-Pollak, and B. A. Kazanskii, Isvest. Nauk. SSSR Otdel Khim Narck 1, 7 (1946).
4. D. W. Mayo and S. K. Freeman, Appl. Spectrosc. 24, 591 (1970).

5. S. K. Freeman and D. W. Mayo, Appl. Spectrosc. 24, 595 (1970).
6. J. N. Willis, Jr., R. B. Cook, and R. Jankow, Anal. Chem. 44, 1228 (1972).
7. R. Obrimsky, Beckman Instruments Co. (private communication, 1972).
8. G. C. Medeiros and G. J. Thomas, Biochem. et Biophys. Acta 247, 449 (1971).
9. K. W. F. Kohlrausch, Zeit. Phys. Chemie B18, 61 (1932).
10. S. Mizushima, *Structure of Molecules and Internal Rotation* (Academic, New York, 1954).
11. C. Altona, Tet. Letters, 2325 (1968).
12. P. Klaboe, Spectrochim. Acta 25A, 1437 (1969).
13. S. K. Freeman, Appl. Spectrosc. 24, 42 (1970).
14. S. K. Freeman, Sci. and Technology, 70 (1967).
15. M. G. J. Beets, Pharm. Reviews 1, 1 (1970).
16. B. Schrader and E. Steigner, Liebigs Ann. Chem. 735, 6 (1970).
17. K. W. F. Kohlrausch, *Der Smekal-Raman-Effect* Ergänzungsband 1931-1937, Struktur und Eigenschaften der Materie XIX (Springer, Berlin).
18. N. B. Colthup, L. H. Daly, and S. E. Wiberly, *Introduction to Infrared and Raman Spectroscopy* (Academic, New York, 1964).
19. R. O. Kagel, Dow Chemical Co. (private communication, 1971).
20. M. P. Groenwege, Spectrochim. Acta 11, 579 (1958).
21. J. H. S. Green, W. Kynaston, and H. M. Parsley, Spectrochim. Acta 19, 549 (1963).
22. R. P. Oertel and D. V. Myhre, Anal. Chem. 44, 1589 (1972).
23. D. W. Mayo and S. K. Freeman, Abstracts Pittsburgh Conference on Analytical Chemistry and Applied Spectroscopy, Cleveland, Ohio, March, 1972, No. 243.
24. G. M. Nakel and L. V. Hayes, J. Agr. Food Chem. 20, 682 (1972).
25. K. Han, Bull. Chem. Soc. Japan 11, 701 (1936).
26. A. R. Katrizky and J. M. Lagowski, J. Chem. Soc., 657 (1959).
27. E. V. Sobolev, V. T. Aleksanyan, R. A. Karakhanov, I. F. Bel'skii, and V. A. Ovodova, Zh. Strukt. Khim 4, (3), 358 (1963).
28. J. J. Peron, P. Saumagne, and M. L. Lebas, C. R. Acad. Sci. Paris Ser. A,B 264B, (10), 797 (1967).
29. N. B. Colthup, J. Opt. Soc. Am. 40, 397 (1950).
30. T. S. Wang and J. M. Sanders, Spectrochim. Acta 15, 1118 (1959).

31. K. H. Michaelian and S. M. Ziegler, Appl. Spectrosc. 27, 13 (1973).
32. C. G. Cannon and G. B. B. M. Sutherland, Spectrochim. Acta 11, 579 (1951).
33. P. Dizabo, H. E. Gatica, N. Le Calvé, G. Franck, J. de Chim. Phys. et de Physiocochemie Biol. 66, 1947 (1969).
34. N. Le Calvé, Ann. Chim. 10, 5 (1965).
35. P. S. Landis, Chem. Rev. 65, 237 (1965).
36. D. Gentric and P. Saumagne, Int. J. Sulfur Chem. 2A, 15 (1972).

Chapter 7

THE ETHYLENE METHYL GROUP

Faced with the problem of ascertaining the presence or absence of an ethylene methyl group, most organic chemists would unhesitatingly select NMR as their spectral tool. There are instances, however, where Raman spectroscopy would be preferable. For example, steric effects occasionally displace the NMR "allylic" methyl signal from its expected range, making it difficult to characterize the group in a compound of unknown structure. Moreover, sample size limitations might preclude the use of NMR while allowing the recording of a Raman spectrum.

Raman spectra of alkanes and methyl cycloalkanes contain very weak bands attributed to methyl symmetric deformations (1365-1390 cm^{-1}) (1), whereas appreciable intensity increase of this mode is observed in olefin spectra when a methyl group is situated on an ethylenic carbon atom. This effect probably arises from a mixing of C=C stretching and CH_3 symmetrical bending vibrations. Moderately intense bands appear at approximately 1375 cm^{-1} in Raman spectra of methyl-substituted aromatics, that is, benzenes, pyridines, pyrazines, pyrimidines, furans, pyrroles, thiazoles, and some thiophenes (2). On the other hand, methyl groups located on saturated as well as unsaturated carbon atoms give rise to moderately strong infrared absorptions in this region, and there is no particular increase in intensity except for acetyl groups. In fact, an IR band near 1350 cm^{-1} in conjunction with one near 1720 cm^{-1} is a good diagnostic for a methyl ketone. The Raman intensities of ethylene methyl symmetric bending vibrations for a large number of acyclic olefins, determined by relating to an internal standard, have been reported to lie between 0.3 and 1.1, whereas 2-methyl-1-alkenes are exceptions (about 0.1) (1). Peak intensity values obtained by internal standard methods require refractive index corrections and, consequently, at least several milligrams of sample must be available.

Raman peak intensity ratios of methyl symmetric deformations at about 1375 cm^{-1} and scattering at about 1440 cm^{-1} of di- and trisubstituted olefins and cycloolefins can be used to diagnose the presence of methyl groups directly attached to a double bonded carbon (2). (The only exceptions encountered so far are 2-methyl-1-alkenes.) In most cases, the number of methyls can be determined. Ratios of 0.26-0.56 were obtained from API spectra (3) of 17 (Z) and (E) acyclic disubstituted compounds containing a single ethylene methyl moiety; values ranging between 0.5 and 0.8 were gained for two such groups in twelve examples of acyclic molecules with trisubstituted double bonds (Table 7.1). Ratios of 0.82-1.0 were observed for three alkenes with three methyls on a double bond (Table 7.1). The significantly lower intensity ratios for molecules where isopropyl, *sec*-butyl, and *tert*-butyl groups are (Z) with reference to the methyl can serve to differentiate between (Z) and (E) isomers. Based on the few examples cited here, it appears that branching on the 1-carbon atom of (Z) substituents may give rise to a steric effect. No significant intensity difference is observed between (Z) and (E) isobutyl isomers.

Pertinent Raman spectral data compiled on 16 acyclic mono-, di-, and trienes in the molecular weight range 96 to 200 are listed in Table 7.2. The dienes and trienes contain an isopropylidene grouping [$=C(CH_3)_2$]. Intensity ratios are in good agreement with the API-derived Raman values (Table 7.1) excepting for the ocimenes. Intensity ratios for 21 mono-, di-, and tricycloalkenes with a single ethylene methyl are presented in Table 7.3. The values for these compounds (0.25-0.50), many of which are natural products, lie in the same range as the acyclic compounds discussed above. The intensity ratio changes brought about by increasing the number of methylene group from two to six in the alcohol residue of angelic acid [2-methyl-(Z)-crotonic acid] are shown in Table 7.4 (4). A summary of the intensity ratios (near 1375 and 1440 cm^{-1}) for the compounds described above appears in Table 7.5.

TABLE 7.1. $I_{1375/1440\ cm^{-1}}$ for Some (Z) and (E) Acyclic Compounds with Ethylene Methyl Groups

Alkene	R	Intensity Ratio
$(H)(H_3C)C{=}C(H)(R)$	C_2 - C_5 i-C_3 t-C_4	0.40 - 0.55 0.29 0.26
$(H)(H_3C)C{=}C(R)(H)$	C_2 - C_4, sec-C_4	0.40 - 0.55
$(H)(H_3C)C{=}C(CH_3)(R)$	C_2, C_3, i-C_4 i-C_3	0.67 - 0.82 0.53
$(H_3C)(H_3C)C{=}C(CH_3)(R)$	C_2, C_3, i-C_3	0.82 - 1.0
$(H_3C)(H_3C)C{=}C(CH_3)(CH_3)$	--	1.5

TABLE 7.2. $I$1375/1440 cm^{-1} for Some Acyclic Mono-, Di-, and Trienes with Ethylene Methyl Groups

Chemical Name*	Structure	Intensity Ratio
3-Methyl-3-hexene *3UY2*	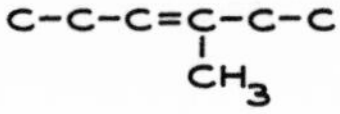 C-C-C=C-C-C CH_3	0.45
4-Hexen-3-one *2V1U2*	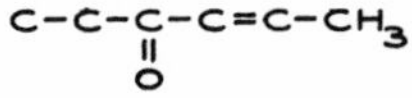 C-C-C-C=C-CH_3 O	0.43
Diethyl methyl maleate *2OVYU1VO2 -C*	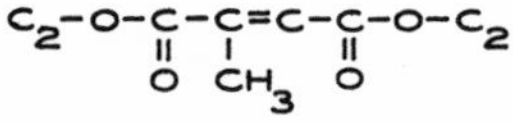 C_2-O-C-C=C-C-O-C_2 O CH_3 O	0.47
3,7-Dimethyl-2-octen-1-ol *Q2UY&3Y*	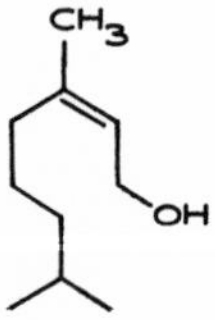 CH_3 OH	0.35
3,7-Dimethyl-2,7-octadien-1-ol *Q2UY&3YU1*	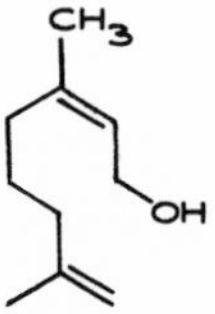 CH_3 OH	0.38
2-Methyl-2-octene *6UY*	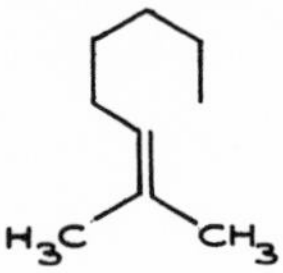 H_3C CH_3	0.65

TABLE 7.2. (Continued)

Chemical Name*	Structure	Intensity Ratio
2,6-Dimethyl-2-octene *2Y&3UY*	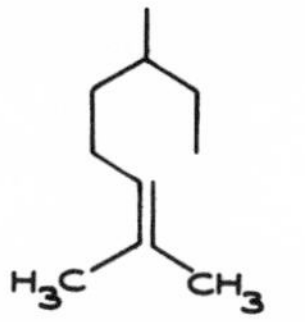	0.60
4,8-Dimethyl-7-nonenal *VH2Y&3Y*	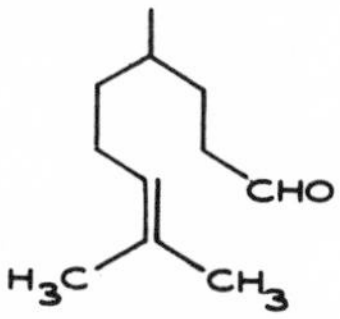	0.53
3,7-Dimethyl-6-octen-1-yl acetate *1Y&U3Y2OV1*	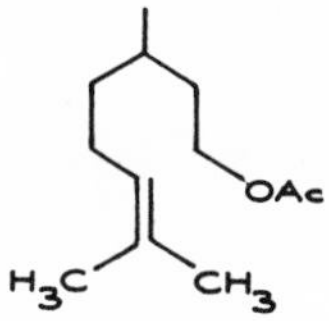	0.70
6-Methyl-5-hepten-2-one *1YU3V1*	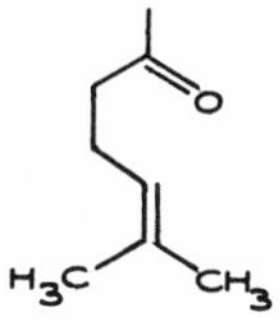	0.75
2-Isopropenyl-5-methyl-4-hexen-1-yl acetate *1Y&U2Y1OV1&YU1*	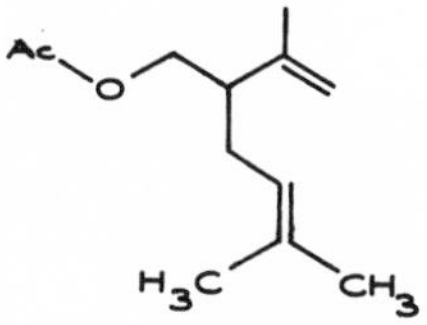	0.75

TABLE 7.2. (Continued)

Chemical Name*	Structure	Intensity Ratio
3,7-Dimethyl-2,6-octadienal *VH1UY&3UY*	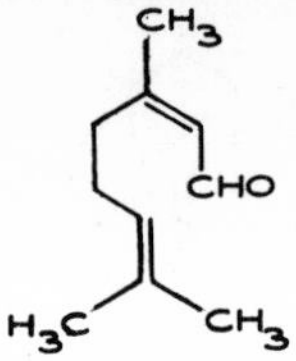	(E) = 1.0 (Z) = 1.0
3,7-Dimethyl-2,6-octadien-1-ol *Q2UY&3UY*	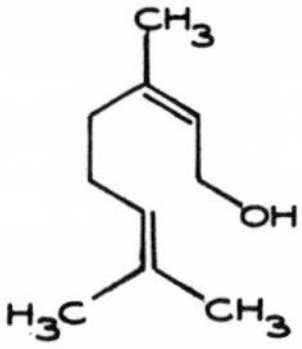	(E) = 1.0 (Z) = 1.0
3,7-Dimethyl-2,6-octadien-1-yl-acetate *1Y&U3Y2U2OV1*	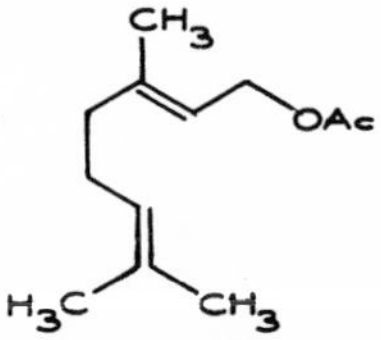	0.9
3,7-Dimethyl-1,3,6-octatriene *1Y&U2UY1U1*	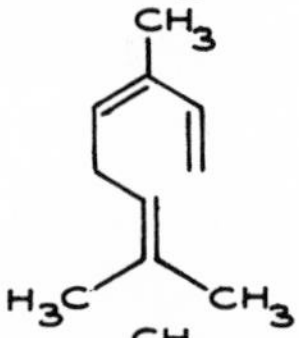	(E) = 0.6 (Z) = 0.6
2,6-Dimethyl-2,4,6-octatriene *2UY&1U2UY*	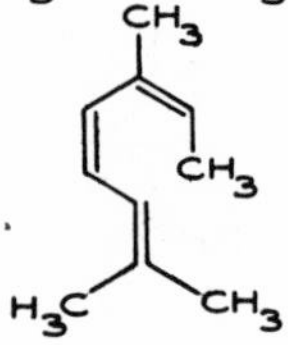	(E) = 1.4 (Z) = 1.6
2,6-Dimethyloctane *2Y&3Y*		<0.1

*Wiswesser Line Notation in italics.

TABLE 7.3. $I$1375/1440 cm^{-1} for Some Mono-, Di-, and Tricycloalkenes with a Single Ethylene Methyl Group

Chemical Name*	Structure	Intensity Ratio
1-Methylcyclopentene *L5UTJ A*	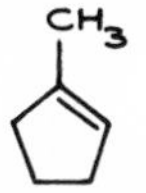	0.25
1-Methylcyclohexene *L6UTJ A*	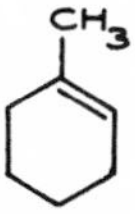	0.50
1-Methylcycloheptene *L7UTJ A*	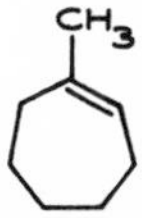	0.26
p-Menth-1-ene *L6UTJ A DY*	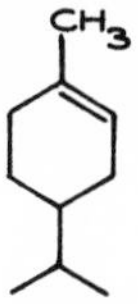	0.40
p-Mentha-1,8-diene *L6UTJ A DYU1*	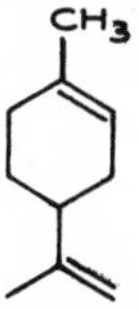	0.40
p-Menth-1-en-8-yl-acetate *L6UTJ A DXOV1*	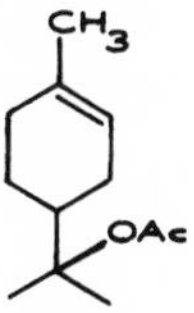 OAc	0.30

TABLE 7.3. (Continued)

Chemical Name*	Structure	Intensity Ratio
p-Menth-1-en-4-ol *L6UTJ A DY DQ*	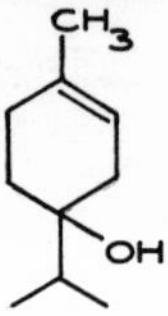	0.40
p-Mentha-1,8-dien-4-ol *L6UTJ A DYU1 DQ*		0.50
2-Hydroxy-p-menth-1-en-3-one *L6V BUTJ BQ C FY*	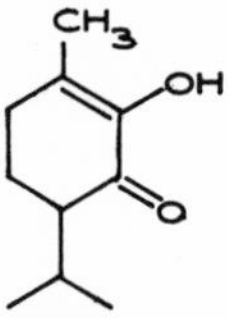	0.30
3,5,5-Trimethyl-2-cyclohexen-1-one *L6V BUTJ C E E*	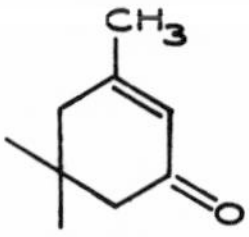	0.30
4,8-Epoxy-p-menth-1-ene *T3OX CHJ C C B-& AL6X CUTJ D*	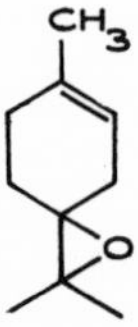	0.40
p-Menthane *L6TJ A DY*	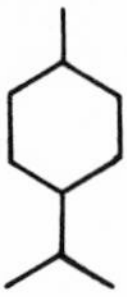	<0.1

TABLE 7.3. (Continued)

Chemical Name*	Structure	Intensity Ratio
2-Carene *L36 DUTJ B B E*	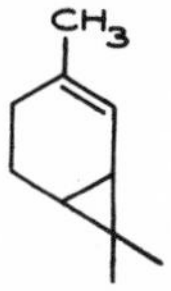 CH_3	0.40
2-Pinene *L46 A EUTJ A A E*	 CH_3	0.30
4,7,7-Trimethyl-6-oxabicyclo [3.2.1]oct-3-ene *T56 A CO GUTJ D D H*	 CH_3, O	0.30
4,11,11-Trimethyl-8-methylene-bicyclo[7.2.0]undec-4-ene (β-Caryophyllene) *L49 EY HUTJ B B EU1 I*	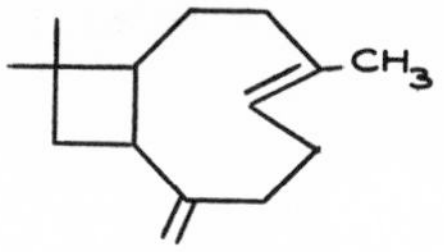 CH_3	0.38
3,3a,4,5,8,8a-Hexahydro-7-isopropyl-2,4-dimethyl azulene *L57 BU HUTJ C F IY*	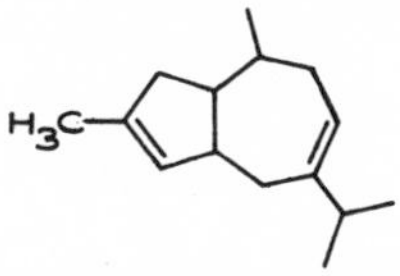 H_3C	0.35
Decahydro-7-isopropenyl-4a-methyl-1-methylene naphthalene *L66 BYTJ BU1 F IYU1*	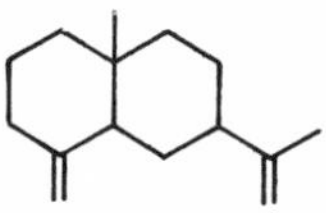	<0.1

TABLE 7.3. (Continued)

Chemical Name*	Structure	Intensity Ratio
1,1a,4,4a,5,6,7,8-Octahydro-1,1,2,4a-tetramethyl cyclopropa[d]naphthalene *L366 A K AX DUTJ B B D G*	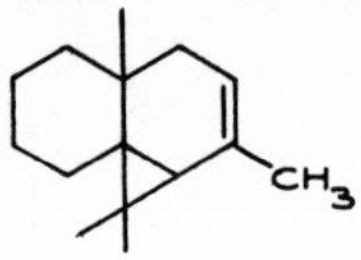 CH_3	0.40
10-Isopropyl-3,7-dimethyl-tricyclo[4.4.0.0^{2,7}]dec-3-ene *L646 A J CUTJ C G JY*	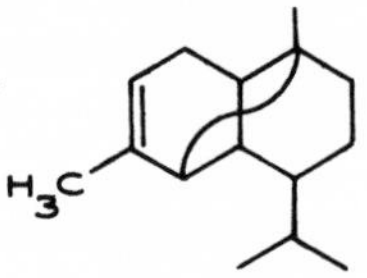 H_3C	0.30
3,3a,4,5,6,7,8,8a-Octahydro-2,4,9,9-tetramethyl-4,7-methanoazulene *L C565 A EUTJ A A B E*	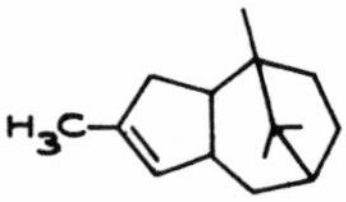 H_3C	0.35
Cedr-8-ene *L65 B5 A 1B BX DUTJ E G G K*	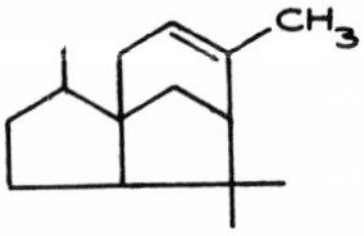 CH_3	0.33
Cedr-8-en-10-one *L65 B5 A 1B BX CV DUTJ E G_ G K*	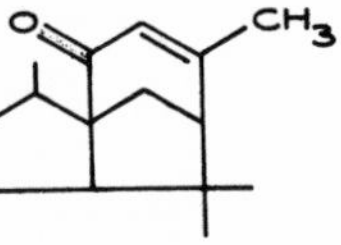 O, CH_3	0.33
Cedrane *L65 B5 A 1B BXTJ E G G K*	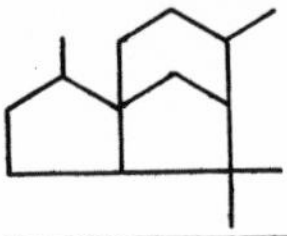	<0.1

*Wiswesser Line Notation in italics

TABLE 7.4. ^{I}1375/1440 cm^{-1} for Some Angelic Acid Esters

$$\begin{array}{ccc} CH_3 & & COOR \\ | & & | \\ C & = & C \\ | & & | \\ H & & CH_3 \end{array}$$

R	Int. Ratio
Ethyl	0.70
Propyl	0.64
Butyl	0.62
Pentyl	0.60
Hexyl	0.57

TABLE 7.5. Summary of ^{I}1375/1440 cm^{-1}

Compound Type	Ethylene Methyls	Int. Ratio 1375/1440 cm^{-1}	Compounds Examined
Acyclic	1	0.26-0.56	21
Cyclic	1	0.25-0.50	20
Acyclic	2	0.50-0.82	17
Acyclic	3	0.82-1.0	9
Acyclic	4	1.3-1.6	3

REFERENCES

1. D. G. Rea, Anal. Chem. 32, 1638 (1960).
2. S. K. Freeman and D. W. Mayo, Appl. Spectros. 26, 543 (1972).
3. American Petroleum Institute Project 44, Chem. Thermo. Properties Center, A & M College of Texas, College Station, Texas.
4. S. K. Freeman (unpublished work).

Chapter 8

ACYCLIC AND ALICYCLIC SULFUR-CONTAINING COMPOUNDS

8.1 INTRODUCTION

Detection of the highly polarizable S-H, C-S, S-S, and S-S-S groups is expeditiously accomplished by Raman spectroscopy, while IR usually is ineffective for all but thiols. The Raman effect is an excellent technique for the examination of natural products containing these moieties. Mercaptans, alkyl sulfides, cyclic sulfides, thiophenes, and thioindanes occur in crude petroleum (1). The sulfate ion in soil and sea water is the source of sulfur for the synthesis of sulfur-containing amino acids found in proteins (methionine, cysteine, and cystine), and sulfur is present in a number of coenzymes (CoA, thioctic acid, and cocarboxylase) (2). The disulfide linkage contributes to the structures of enzymes (3) and hormones (4). Insulin (see p.272) and the two polypeptide hormones of the posterior lobe of the hypophysis, pitressin and oxytocin, are inactivated by reduction of their S-S bridges (2). A host of organosulfur compounds are found in nature (1,3,5), many of which are valuable in the flavor and perfumery industries (6).

Reports concerned with the applications of Raman spectroscopy to the examination of organosulfur compounds began to appear in the literature more than forty years ago (7). Prior to 1968 numerous investigators had published accounts of their studies in this field (see, e.g., References 8-18).

Conformational details of sulfur-containing molecules often can be obtained from a study of their Raman spectra in liquid and solid states. Combined mass and Raman spectral data serve as powerful structural clues, especially in natural product studies where sample size limitations may prevent generation of NMR spectra. The number of sulfur atoms present in a molecule can be calculated from a low resolution mass spectrum employing the ^{34}S isotope. Since the abundance of this isotope is about 5%, the ratio of the molecular ion and the M+2 ion is approximately 20/1 for one sulfur atom, 10/1 for two sulfur atoms, and so on. Raman spectroscopy allows differentiation among thiols, mono-, di-, and trisulfides; more polar linkages such as C=S, S=O, SO_2, SO_3H, and so on, give rise to strong IR absorptions (19). Thus, a confident statement can be made concerning the type of sulfur bonding in a molecule from its vibrational spectra. Obviously, the degree of unsaturation, other functionalities, molecular weight, and fragmentation pattern are valuable aids in gaining additional structural information.

8.2 THIOLS

Spectral data for various thiols are listed in Table 8.1 (20). A strong Raman band at 2560±20 cm^{-1} ($\rho \sim 0.1$) is nearly incontrovertible evidence for a thiol. In addition to this S-H stretching frequency, a moderate to strong C-S stretching band is observed between about 625-725 cm^{-1} (ρ = 0.06-0.14). For thiols capable of existing as *trans* and *gauche* conformers with respect to the C-CS linkage, a weak band is displayed near 705 cm^{-1} attributed to the more stable *trans* rotamer while a strong band for the *gauche* form appears at about 650 cm^{-1}. In the case of 1-propanethiol, the intensity of the 703 cm^{-1} band disappears when going from the liquid to the solid state (21). Accordingly, the Raman scattering for 1-butanethiol (708 cm^{-1}), 1-pentanethiol (700 cm^{-1}), 1-heptanethiol (703 cm^{-1}), 1-decanethiol (702 cm^{-1}), 3-methyl-1-butanethiol (708 cm^{-1}), and 1,3-propanedithiol (699 cm^{-1}) may be assigned to the *trans* form. The depolarization ratios of the C-S stretching bands for *trans* conformers ($\rho \sim 0.25$) are significantly higher than for corresponding *gauche* forms ($\rho \sim 0.12$) (20).

8.3 SULFIDES

TABLE 8.1. Raman Frequency and Depolarization Value Data for the S-H and C-S Stretching Vibrations of Some Thiols (20)

Chemical Name*	Structure	ν_{S-H} (cm^{-1})	ρ	ν_{C-S} (cm^{-1})	ρ
Ethanethiol *SH2*	C–C–SH	2572[b]	0.06	659[a]	0.10
1,2-Ethanedithiol *SH2SH*	HS–C–C–SH	2560[a]	0.09	633[b]	0.06
2-Mercaptoethanol *SH2Q*	HO–C–C–SH	2567[a]	0.08	662[b]	0.10
2-Mercaptoacetic Acid *SH1VQ*	HOOC–C–SH	2565[b]	0.11	670[a]	0.14
3-Mercaptopropionic Acid *SH2VQ*	HOOC–C–C–SH	2570[b]	0.11	670[b]	0.14
1-Propanethiol *SH3*	C–C–C–SH	2572[c]	0.08	703 650[a]	0.09 0.12
2-Propene-1-thiol *SH2U1*	C=C–C–SH	2578	0.1	750 sh 721	0.10
3-Mercapto-1,2-propanediol *SH1YQ1Q*	HO–C–C(OH)–C–SH	2569[b]	0.10	679[d]	0.09
L-Cysteine HCl *SHYZVQ GH*	HOOC–C($NH_3^+Cl^-$)–SH	-	-	686[a]	-
4-Mercapto-2-butanone *SH2V1*	C–C(=O)–C–C–SH	2568[b]	0.09	653[a]	0.14
3-Mercapto-2-butanone *SHYV1*	C–C(=O)–C(SH)–C	2570	0.09	678 660[a]	0.11
1,3-Propanedithiol *SH3SH*	HS–C–C–C–SH	2562[b]	0.08	699 655[a]	∿0.25 0.08

TABLE 8.1. (Continued)

Chemical Name*	Structure	ν_{S-H} (cm^{-1})	ρ	ν_{C-S} (cm^{-1})	ρ
1-Methyl-1-propanethiol *SHY2*	C-C-C(C)-SH	2572	0.11	6.25[a]	0.11
1,2-Dimethyl-1-propanethiol *SHY&Y*	C-C(C)-C(C)-SH	2570	0.10	618 642[a] 678	0.1 0.09 0.13
2-Methyl-1-propanethiol *SH1Y*	C-C(C)-C-SH	2574	0.09	670[wk] 712[a]	0.04 0.11
1-Butanethiol *SH4*	C-C-C-C-SH	2575[b]	0.09	708[e] 658[a]	0.21 0.11
2-Methyl-1-butanethiol *SH1Y2*	C-C-C(C)-C-SH	2578	0.09	668	0.11
2-Methyl-2-propanethiol *SHX*	C-C(C)(C)-SH	2570[c]	0.11	590[a]	0.07
3-Methyl-1-butanethiol *SH2Y*	C-C(C)-C-C-SH	2575[b]	0.10	708 658[a]	∿0.25 0.11
1-Pentanethiol *SH5*	C-C-C-C-C-SH	2576[b]	0.10	700[e] 651[a]	0.25 0.11
1-Heptanethiol *SH7*	n-C_7SH	2575[c]	0.10	703[e] 655[a]	0.25 0.12
1-Nonanethiol *SH9*	n-C_9SH	2574[c]	0.08	-	-
1-Decanethiol *SH10*	n-C_{10}SH	2575[c]	0.09	702 653[a]	∿0.25 0.14
α-Toluenethiol *SH1R*	ϕ-C-SH	2575	0.14	679[b]	0.08

TABLE 8.1. (Continued)

Chemical Name*	Structure	ν_{S-H} cm^{-1}	ρ	ν_{C-S} cm^{-1}	ρ
Benzenethiol *SHR*	ø-SH	2570	0.10	698[d]	0.07
m-Toluenethiol *SHR C*	C, SH	2572	0.10	687[b]	0.07
2-Furan methanethiol *T5OJ B1SH*	O, CSH	2572	0.10	602 (?)	0.09
Cyclopentanethiol *L5TJ ASH*	SH	2568	0.07	-	-
Cyclohexanethiol *L6TJ ASH*	SH	2570	0.08	-	-

*Wiswesser Line Notation in italics

[a]Strongest band below 2800 cm^{-1}.

[b]2nd Strongest band below 2800 cm^{-1}.

[c]3rd Strongest band below 2800 cm^{-1}.

[d]4th Strongest band below 2800 cm^{-1}.

[e]trans Rotamer (probably).

[wk]Weak band.

8.3.1 Acyclic Sulfides

Considerable structural information concerning an acyclic sulfide often can be gained from the Raman band profile between 600 and 725 cm^{-1} and depolarization ratios. Comparison of liquid and solid state spectra yields insights concerning conformational states.

The CH_3-S Group Frequency

An examination of the spectra of a series of linear alkyl methyl sulfides (Table 8.2) reveals that after the alkyl chain length has reached approximately five carbons the asymmetric C-S-C stretching mode of the *trans* conformer becomes constant (720 ± 1 cm^{-1}) (22). Furthermore, this intense mode remains relatively unaffected by alkyl substitution on the alkyl chain, with an average value for the stretching frequency of 722 ± 7 cm^{-1}. When the methylene group is replaced by a large hetero atom such as sulfur attached directly to the sulfur atom, a significant disruption of mechanical coupling of these modes and a subsequent drop of approximately 30-40 cm^{-1} in frequency would be expected. A frequency shift is observed for thiomethyl sulfides (Table 8.3) in which the average value of the C-S stretch is reduced 29 cm^{-1}. This observation may be interpreted as indicating that these thiomethyl compounds possess a simple and essentially unperturbed C-S stretching frequency in the 693 ± 5 cm^{-1} region. The tight wavenumber range, consistent depolarization ratios (Tables 8.2 and 8.3), and high band intensities result in this Raman skeletal group frequency being of particular value in the identification of the presence of the CH_3-S group. Raman spectra of methyl sulfoxide, sulfone, diimine, and imine sulfoxide show strong, polarized bands near 672-700 cm^{-1} (18,23).

Methyl Alkyl Sulfides

When the alkyl residue is propyl or higher, a characteristic series of bands is observed in the C-S stretching region (Table 8.4) (20). Tentative assignments of these bands to *trans* and *gauche* conformers, based on data for ethyl methyl sulfide, depolarization measurements, and low temperature studies, are presented in Table 8.5 (20,23).

TABLE 8.2. CH_3-S Stretching Frequencies in Mono-Sulfides (CH_3-S-R) (22)

R*	ν_{CH_3-S} (cm^{-1})	ρ
H *SH1*	709	0.04
CH_3 *1S1*	695	0.04
C_2H_5 *2S1*	727	0.20
C_3H_7 *3S1*	725	0.25
i-C_3H_7 *1YS1*	726	0.22
C_4H_9 *4S1*	724	0.25
sec-C_4H_9 *2YS1*	721	0.26
C_5H_{11} *5S1*	720	0.26
i-C_5H_{11} *1Y2S1*	712	0.25
C_7H_{13} *7S1*	719	0.25
C_8H_{15} *8S1*	720	0.25
$C_{10}H_{19}$ *1ØS1*	721	0.24
C-S-C *1S1S1*	722	0.21
C-S-C(-C) *1SYS1*	728	0.22
C-C-C-CHO *VH3S1*	720	-
C-C-C-COOH *QV3S1*	726	-
C-C(NH_2)COOH *QVYZ1S1*	716	-
C-C-$COOC_3$ *3OV2S1*	726	0.34
C-C-C(=O)-S-C *1SV2S1*	720	0.23
C(=O)-C(-C)-C *1YVS1*	725	0.28
C-ϕ *1S1R*	700	0.10
C(=O)-ϕ *1SVR*	726	0.30
ϕ *1SR*	693	0.09
ϕ(4-OH) *QR DS1*	719	-
2-Furyl *T5OJ BS1*	705	0.23
C-C=C *1U2S1*	705	0.06

*Wiswesser Line Notation in italics.

TABLE 8.3. CH_3-S Stretching Frequencies in Di- and Trisulfides (CH_3-S-R) (22)

R	ν_{CH_3-S} (cm^{-1})	ρ
S-C	691	0.15
S-C_3	695	0.15
S-C_3^i	692	0.15
S-C_4^t	688	0.19
S-Cϕ	692	0.17
S-C=C-C	694	0.13
S-S-C	697	0.15

Band patterns for short branched-chain derivatives such as isopropyl methyl and isobutyl methyl sulfides differ considerably from their long chain relatives (Figure 8.1). When a methyl group is located on the 3-carbon atom or higher, as in isopentyl methyl sulfide, the scattering pattern closely resembles that of the methyl linear alkyl sulfides. The Raman bands associated with the C-S stretching vibrations of isobutyl methyl sulfide (Figure 8.1) have been tentatively assigned with the aid of depolarization measurements and comparison with spectra of the chloride and thiol derivatives (20). Solely on the basis of band location it would be difficult to assign the strong 702 cm^{-1} and moderately intense 718 cm^{-1} bands. However, the depolarization ratios permit one to ascribe the former ($\rho = 0.06$) to the CH_2-S stretch and the latter ($\rho = 0.17$) to the CH_3-S asymmetric stretch. A weak band at 671 cm^{-1}, which disappears in the solid state spectrum, probably originates in the higher-energy CH_2-S rotamer and therefore is the same vibrational mode as the 670 cm^{-1} line in the thiol (Table 8.1, 2-methyl-1-propanethiol).

TABLE 8.4. Raman Spectral Data for Some Methyl Alkyl and Methyl Branched Alkyl Sulfides (20,23)

Sulfide*	Structure	ν_{C-S} (cm^{-1})	ρ
Methyl *1S1*	C-S-C	745[c] 695[a]	0.75 0.04
Ethyl methyl *2S1*	C-C-S-C	727[c] 675[b] 656[a]	0.20 0.07 0.04
Methyl propyl *3S1*	C-C-C-S-C	725[c] 700[a] 672[d] 650[b]	0.25 0.10 0.15 0.10
Isopropyl methyl *1YS1*	C-C(C)-S-C	726[b] 640[a] 612[e]	0.22 0.07 0.04
Butyl methyl *4S1*	n-C_4-S-C	724[b] 700[a] 673[e] 651[c]	0.25 0.08 0.08 0.07
Isobutyl methyl *1Y1S1*	C-C(C)-C-S-C	718[b] 702[a] 671 wk	0.17 0.06 -
Methyl pentyl *5S1*	n-C_5-S-C	720[b] 695[a] 669[d] 650[c]	0.26 0.08 0.07 0.07
Isopentyl methyl *1Y2S1*	C-C(C)-C-C-S-C	712[e] 697[a] 675[w] 650[c]	0.25 0.06 0.10 0.05

*Wiswesser Line Notation in italics

[a-e]Band intensities below 1400 cm^{-1} ranked in decreasing order; a = strongest.

Alkyl Sulfides Other Than Methyl

The pre-laser Raman spectrum of liquid ethyl sulfide was interpreted to signify the presence of three conformers (24). Three bands between 600 and 700 cm^{-1} were assigned to the TT, TG, and GG forms in order of decreasing frequency. The two low frequency bands present in the liquid disappeared in the solid, demonstrating that the most stable conformer was the one whose C-S stretching vibration appeared at 693 cm^{-1} (Figure 8.2).

Replacing one or two of the ethyl groups in ethyl sulfide with propyl or higher residues results in the appearance of a complex scattering region between 600 and 700 cm^{-1} (Figure 8.2, Table 8.6) (20).

TABLE 8.5. Assignments for CH_3-S Stretching Bands of CH_3SR (20,23)

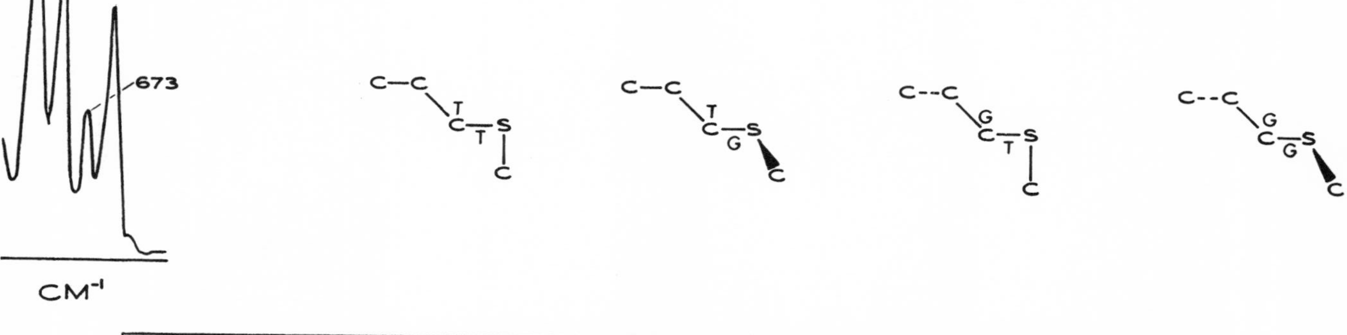

Band	C-C-S-C		RC-C-C-S-C*		Assignment (Tentative)	Comments
	cm^{-1}	ρ	cm^{-1}	ρ		
1	727	0.20	722±3	0.25±0.01	CH_3-S (T) asym. +C-CS (G)	2 bands, poorly resolved
2	Missing		696±4	0.09±0.01	C-CS (T)	Rel. intensity >on cooling
3	675	0.07	670±3	0.11±0.04	CH_3-S (G)	Rel. intensity <on cooling
4	656	0.04	650±1	0.09±0.02	CH_3-S (T & G) sym.	-

*R=H, CH_3, or homologues

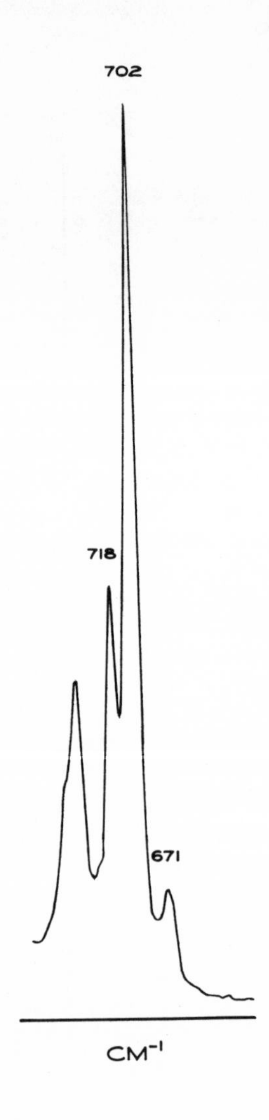

Figure 8.1. Raman spectrum of isobutyl methyl sulfide.

Scanning the 600-700 cm^{-1} region at slower speeds with increased spectral resolution partially separates the shoulders, and cooling of the sample sharpens them to a considerable extent. These bands probably originate from several rotamer types. Since no significant changes occur in relative band intensities on cooling,

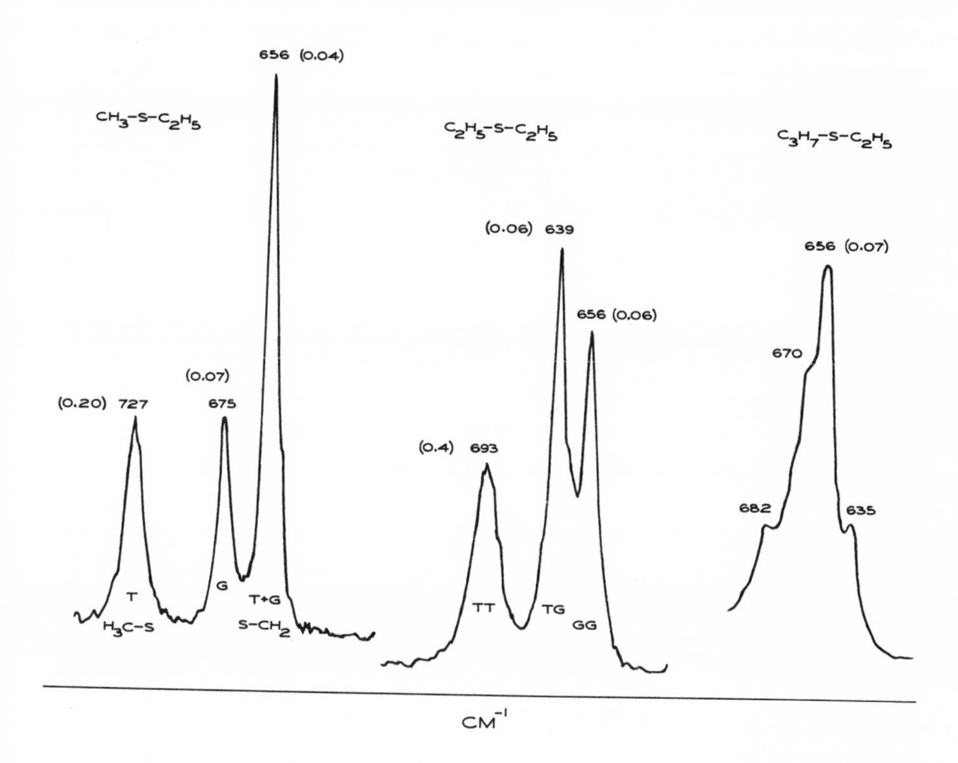

Figure 8.2. Raman spectra of some methyl alkyl sulfides (20,23).

the conformers are of equal energy in the liquid phase. The depolarization values of the central, major band at about 650 cm^{-1} span the narrow range 0.07-0.10 and three shoulders are seen at about 635, 670, and 685 cm^{-1}.

Similar to the 1- and 2-methyl-substituted methyl alkyl sulfides, isopropyl, isobutyl, and *sec*-butyl sulfides display significantly different scattering patterns from their linear analogues between 600 and 725 cm^{-1} (20). The spectrum of isopentyl sulfide, on the other hand, as observed in the methyl alkyl series, resembles the straight-chain derivatives.

TABLE 8.6. Raman Spectral Data for Some Alkyl Sulfides Other Than Methyl (20)

Sulfide*	Structure	ν (cm^{-1}) Major band	ρ	Band "Shoulder"
Ethyl *2S2*	C-C-S-C-C	693 639 656	0.4 0.06 0.06	- - -
Ethyl propyl *3S2*	C-C-S-C-C-C	656	0.07	682 670 635
Propyl *3S3*	C-C-C-S-C-C-C	646	0.07	678 ~665 630
Butyl *4S4*	$n-C_4-S-C_4-n$	655	0.09	688 668 635
Butyl ethyl *4S2*	$n-C_4-S-C-C$	652	0.09	685 670 635
Ethyl heptyl *7S2*	$n-C_7-S-C-C$	657	0.08	688 675 640
Decyl ethyl *10S2*	$n-C_{10}-S-C-C$	657	0.07	690 670 638

*Wiswesser Line Notation in italics

Allyl Sulfides

The 735 cm^{-1} band in the pre-laser Raman spectrum of allyl sulfide was attributed to the allyl-S stretch of the "twisted *trans*" rotamer (25). Recently, this vibrational mode in alkyl, alkenyl, and allyl phenyl sulfides has been assigned to the 750 cm^{-1} band (26) (Figure 8.3, Table 8.7).

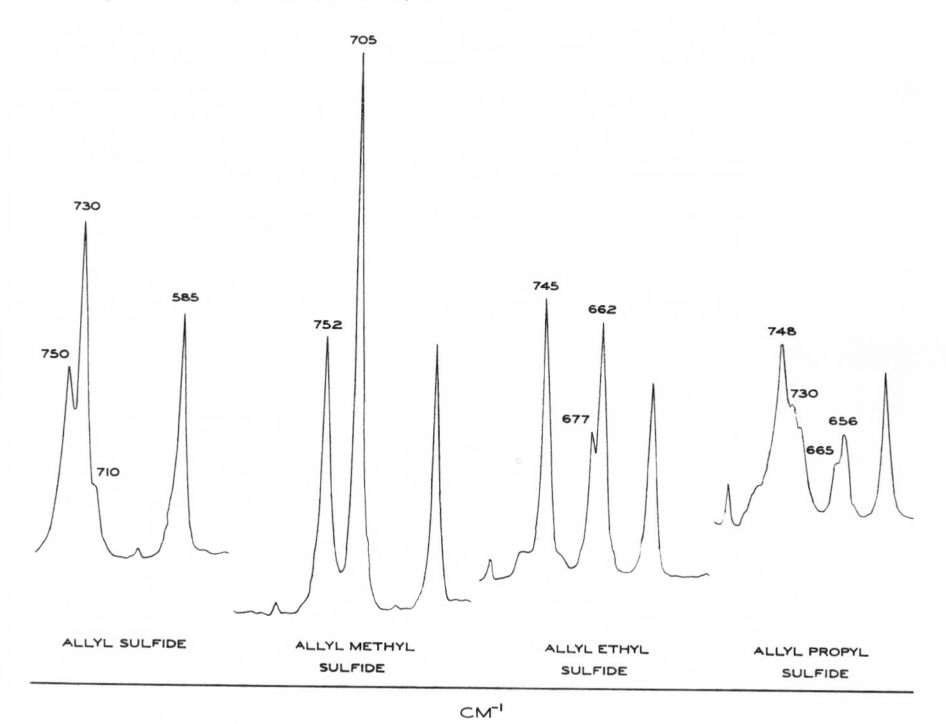

Figure 8.3. Raman spectra of some allyl sulfides (26).

Conformational isomerism, not present in allyl methyl sulfide, is manifested in the spectrum of allyl ethyl sulfide by two bands at 662 and 677 cm^{-1} arising from hindered rotation about the C-S bond. The high frequency band disappears in the solid state spectrum and therefore is ascribed to the CH_2-S stretch of the less stable conformer. Two poorly resolved bands observed at 720 and 730 cm^{-1} in the spectrum of allyl propyl sulfide, absent in the ethyl analogue, may be due to rotamers about the C-CS bond.

TABLE 8.7. Raman Spectral Data for Some Allyl Sulfides (26)

Sulfide*	Structure	ν_{C-S} (cm^{-1})	ρ
Allyl *1U2S2U1*	C=C–C–S–C–C=C	748 730 710 sh	0.30 0.10
Allyl methyl *1U2S1*	C=C–C–S–C	752 705	0.45 0.06
Allyl ethyl *2S2U1*	C=C–C–S–C–C	745 677 662	0.2 0.11 0.06
Allyl propyl *3S2U1*	C=C–C–S–C–C–C	748 730 sh 720 sh 665 652	0.3 - - 0.10 0.10
Allyl isopropyl *1YS2U1*	C=C–C–S–C(–C)–C	745 700 sh 639	0.25 - 0.10
Allyl sec butyl *2YS2U1*	C=C–C–S–C(–C)–C–C	748 688 649 628	0.30 - - -
Allyl phenyl *1U2SR*	C=C–C–S–Φ	738 699	0.5 0.11

*Wiswesser Line Notation in italics

The tert-Butyl-Sulfur Bond

The C-S symmetric stretching band of *tert*-butyl thiol is observed at an unusually low frequency (Table 8.8 - 590 cm^{-1}, ρ = 0.07). A weak band at 608 cm^{-1} (ρ = 0.5), may originate from the asymmetric stretch. *tert*-Butyl chloride shows a band at 569 cm^{-1} (ρ = 0.09). In agreement with these data, *tert*-butyl sulfide displays a 582 cm^{-1} band (ρ = 0.04) and the spectrum of *tert*-butyl propyl sulfide contains its strongest band at 599 cm^{-1}. Conformers about the propyl-sulfur bond give rise to C-S stretching bands at 682 cm^{-1} (ρ = 0.12) and 663 cm^{-1}. It is quite likely, then, that the stretching frequency for *tert*-butyl-S lies between about 580 and 600 cm^{-1} (20)

TABLE 8.8. Raman C-S Stretching Frequencies for Some *tert*-Butyl Compounds (20)

tert-Butyl Compound	ν_{C-S} (cm^{-1})	ρ
Chloride	569	0.09
Thiol	590	0.07
Sulfide	582	0.04
Propyl sulfide	599	0.07
	663	∿0.1
	682	0.12

This conclusion is important for the discussion concerned with the S-S stretching frequency in *tert*-butyl compounds (see p.229).

The Phenyl-Sulfur Bond

The C-S stretching mode involves three different vibrational types which are depicted below. These are referred to as "X-sensitive", where X signifies a substituent. Vibrations *r* and *t* generally give rise to moderate-strong Raman bands. The *r* band, observed at about 700 cm^{-1} (ρ ∿0.1) for the phenyl-S stretch, is the one assigned for compounds containing a sulfur directly attached to the phenyl ring, that is, phenyl thiol, allyl phenyl sulfide, and phenyl disulfide.

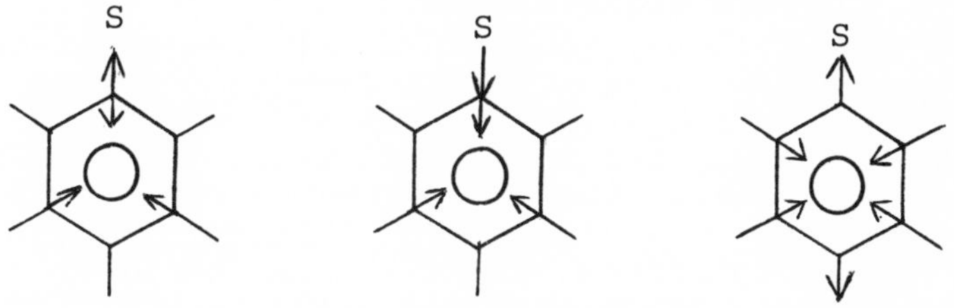

S S S

q *r* *t*

8.3.2 Cyclic Sulfides

The C-S stretching vibration in cyclic sulfides usually appears as a strong band between 600 and 720 cm^{-1} (ρ = 0.05-0.15). Groups of bands arising from geometrical isomers and/or conformers often are observed in this region. A perusal of Table 8.9 makes it clear that with but few exceptions, neither ring size nor presence of additional sulfur atoms or other hetero atoms noticeably affects the C-S stretching frequency (20). Consequently, the absence of a moderate to intense band at 600-720 cm^{-1} usually is good evidence for the absence of a cyclic sulfide. When a molecule contains a very strong scattering moiety such as a cyclohexyl or a phenyl ring, the C-S band intensity becomes moderate to weak, for example, 2-phenyl thiepane. Similar to *tert*-butyl sulfide (582 cm^{-1}), steric hindrance appears to reduce the C-S stretching frequency. When a sulfur atom is flanked by tetrasubstituted carbons, a band tentatively assigned to this C-S mode appears near 510-550 cm^{-1}.

Russian workers (27) have reported that the overall intensity of the bands between 600 and 740 cm^{-1}, attributed to the C-S stretch in a variety of cyclic sulfides, is constant. Exceptions to this finding occur when the C-S group is located next to a double bond or an aromatic system. In these instances the total intensity is very low.

8.4 DISULFIDES

8.4.1 Acyclic Disulfides

Two strong Raman bands are manifested in Raman spectra of acyclic disulfides: C-S stretch (about 600-725 cm^{-1}, $\rho \sim 0.15$) and S-S stretch (about 450-550 cm^{-1}, $\rho \sim 0.10$). Acyclic molecules containing the disulfide linkage can adopt at least two conformations that are optical antipodes. Because the charge density on the sulfur atoms is unsymmetrical about the bond axis, mutual repulsion between the unshared pairs of $3p\pi$ electrons on adjacent sulfur atoms is an important factor in hindering free rotation. As a result, the most stable conformer is the one with a dihedral angle of about 90° between substituents.

TABLE 8.9. Raman Spectral Data for Some Cyclic Sulfides (20)

Chemical Name*	Structure	ν_{C-S} (cm^{-1})	ρ	Rel. Int.
Thiirane *T3STJ*		616	0.09	10
Methyl Thiirane *T3STJ B*		602	0.16	10
2,3-Epithiopropyl methyl ether *T3STJ B1O1*	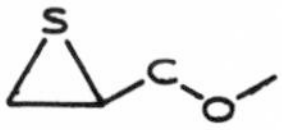	596	0.10	10
Thietane *T4STJ*		698	0.1	6
2-Methyl thietane *T4STJ B*		630 700	- -	- -
3-Ethyl-2-propyl-2H-thiete *T4S AHJ B3 C2*	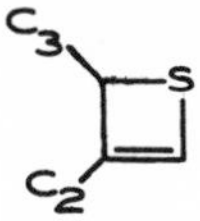	710	-	-
Tetrahydrothiophene *T5STJ*		688	0.08	10
Tetrahydro-2-methyl-thiophene *T5STJ B*	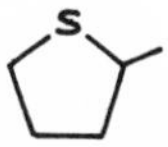	683 659 630	0.1 0.08 0.09	10 5 7
Tetrahydro-3-methyl-thiophene *T5STJ C*		683	-	-
Tetrahydro-2-phenyl-thiophene *T5STJ BR*	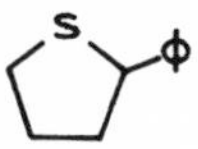	685	0.11	3
2,5-Dihydrothiophene *T5S CUTJ*		710	-	10

TABLE 8.9. (Continued)

Chemical Name*	Structure	ν_{C-S} (cm^{-1})	ρ	Rel. Int.[a]
Dihydro-2(3H)-thiophenone (γ-Thiobutyrolactone) *T5SVTJ*	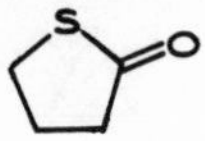	632 688	0.03 0.05	10 9
Dihydro-3(2H)-thiophenone *T5S CVTJ*	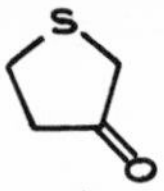	680	0.10	5
5-Methyl dihydro-2(3H)-thiophenethione *T5SYTJ BUS E*	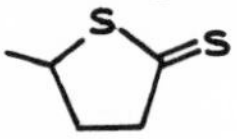	640	0.10	10
Thiazolidine *T5M CSTJ*	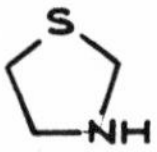	702 672	0.13 0.07	7 7
2,4-Thiazolidinedione *T5SVMV EHJ*	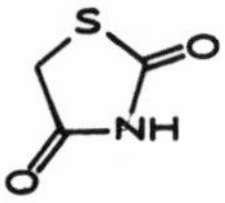	614	-	10
Tetrahydro-2H-thiopyran *T6STJ*		655	0.05	10
2-Methyl tetrahydro-2H-thiopyran *T6STJ B*	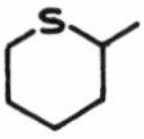	645	0.1	10
2-Phenyl tetrahydro-2H-thiopyran *T6STJ BR*	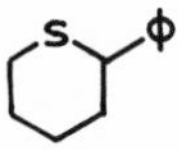	675	-	7
Thiepane *T7STJ*	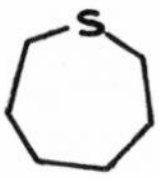	652	10	0.05
2-Phenyl thiepane *T7STJ BR*	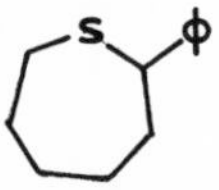	674 665	0.15 0.11	3 3
1,4-Oxathiane *T60 DSTJ*	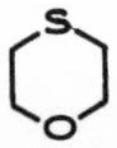	688	0.05	10

TABLE 8.9. (Continued)

Chemical Name*	Structure	ν_{C-S} (cm^{-1})	ρ	Rel. Int.[a]
3-Methyl-1,4-Oxathiane *T60 DSTJ C*	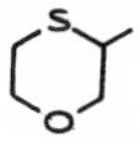	650	0.1	10
3-Phenyl-1,4-Oxathiane *T60 DSTJ CR*	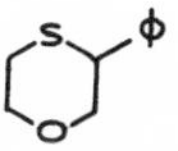	683	0.18	10
1,4,9,13-Tetraoxa-6,17-dithiacyclononadecane *T-19-S CO FO IS MO QOTJ*	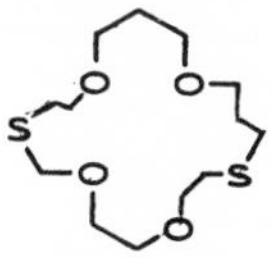	662	-	10
2,8,15,21-Tetrathia-5,18-dioxatricyclo [20,4,0,0^{9,14}] hexacosane *T I6-18-6 BS EO HS OS RO_ USTJ*	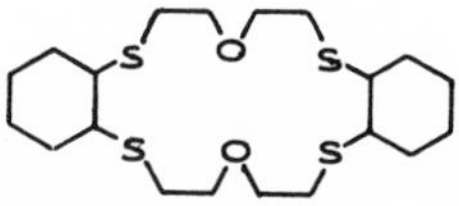	680 710	- -	5 10
1,3-Dithiolane *T5S CSTJ*		698 685	0.10 0.07	5 10
2-Methyl-1,3-dithiolane *T5S CSTJ B*	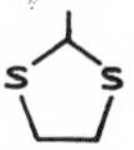	679 628	0.15 0.07	9 10
4-Methyl-1,3-dithiolane *T5S CSTJ D*	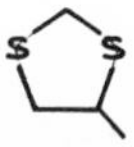	692 640	0.08 0.08	10 8
2,4-Dimethyl-1,3-dithiolane *T5S CSTJ B D*	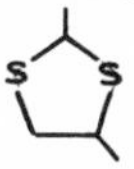	628	0.04	10
$\Delta^{2,2'}$-Bi-1,3-dithiolane *T5SYSTJ B- 2U*	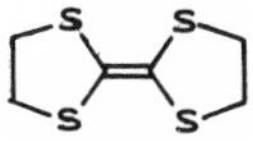	679	10	-
2,2'-Methylene bis(2-methyl-1,3-dithiolane) *T5SXSTJ B-& 2 A DL4X CXTJ*	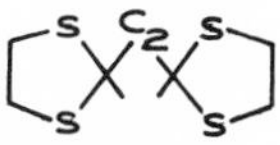	650	-	10

The barrier to free rotation about the S-S bond is approximately 10 kcal/mole (28). Although this type of conformational information has not been obtained from vibrational spectra, it is possible to detect rotamers

TABLE 8.9. (Continued)

Chemical Name*	Structure	C-S (cm^{-1})	ρ	Rel. Int.[a]
2,4-Dimethyl-1,3-dithiolane-2-carboxaldehyde *T5S CSTJ BVH B D*	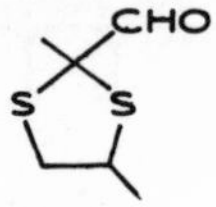	640	-	10
1,3-Dithiolan-2-yl-methyl ketone *T5S CSTJ BV1*	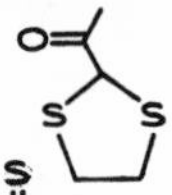	678	-	10
1,3-Dithiolane-2-thione *T5SYSTJ BUS*	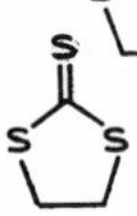	678	-	5
m-Dithiane *T6S CSTJ*		642	-	10
2-Methyl-m-dithiane *T6S CSTJ B*	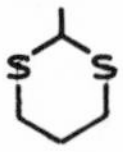	627	0.08	10
4-Methyl-1,3-dithiolane *T5S CSTJ D*	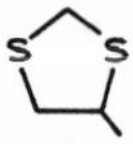	628	0.08	10
2,4-Dimethyl-1,3-dithiolane *T5S CSTJ B D*	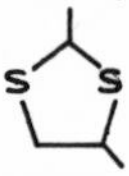	625	0.07	10
2,4-Dimethyl-m-dithiane *T6S CSTJ B D*	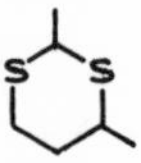	637	0.07	10
2,4,6-Trimethyl-m-dithiane *T6S CSTJ B D F*	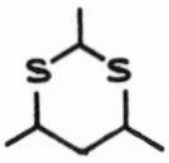	631 620 592	0.03 0.11 0.11	10 10 7
p-Dithiane *T6S DSTJ*		631	-	10
s-Trithiane *T6S CS ESTJ*	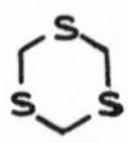	651	-	10

originating in hindered rotation about the $S-C_a$ and C_a-C_b bonds in $R-S-S-C_a-C_b-C_c$. Conformers about the C_b-C_c, and other bonds, are spectroscopically indistinguishable.

TABLE 8.9. (Continued)

Chemical Name*	Structure	ν_{C-S} (cm^{-1})	ρ	Rel. Int.[a]
trans-2,4,6-Trimethyl-s-trithiane *T6S CS ESTJ B D F -T*	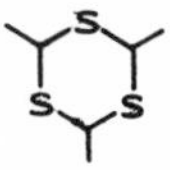	592	-	10
cis-2,4,6-Trimethyl-s-trithiane *T6S CS ESTJ B D F -C*	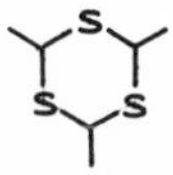	645	-	10
Hexamethyl-s-trithiane (Trithioacetone) *T6S CS ESTJ B B D D F F*		513	-	10
5-Thia-10,11-diazadispiro [3.1.3.2]undec-10-ene *T5NNXSXJ C-& E-&/ AL4XTJ 2*	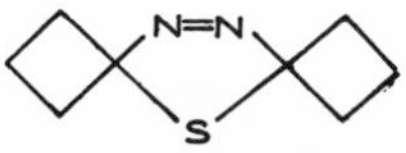	555	-	10
5-Thia-10,11-diazadispiro [3.1.3.2]undecane *T5MMXSXJ C-& E-&/ AL4XTJ 2*	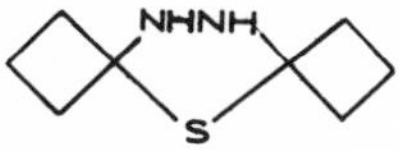	547	-	10
7-Thiabicyclo [2,2.1] heptane (7-Thianorbornane) *T55 A ASTJ*		600 565	- -	3 8
9-Thiabicyclo [3.3.1] 2-nonene *T66 A AS CUTJ*		638	-	-
2,4-Diadamantyl-1,3-dithietane *L66 B6 A B- C 1B ITJ B- 2 B D T4S CSTJ*	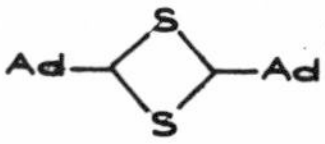 Ad–(S,S dithietane)–Ad [b]	646	-	5
2,4,6-Triadamantyl-s-trithiane *L66 B6 A B- C 1B ITJ B- 3 B D FT6S CS ESTJ*	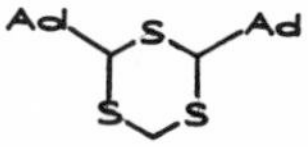	668	-	5

*Wiswesser Line Notation in italics

[a]The value of 10 signifies the strongest band between 300-1400 cm^{-1}. Numbers 9,8,7, etc. denote bands whose intensities are 90%, 80%, 70%, etc. of the strongest band.

[b]Ad=Adamantane

The C-S-S-C group should give rise to two S-S stretching vibrations, namely, symmetric and asymmetric. If the dihedral angle around the S-S bond is not 180°, both are Raman active, but the symmetric and asymmetric frequencies are essentially equal if the dihedral angle is approximately 90° (29). The appearance of two closely lying bands near 500 cm^{-1} might be ascribed to these two vibrational modes but depolarization ratios and low temperature spectra can serve to distinguish them from rotamer bands (30).

Disulfides may be formed in small amounts by air oxidation of thiols. The extent of this reaction can be ascertained from a standard curve using the peak height ratios of C-S and S-S stretching bands. For example, as little as about 0.1% cystine can be determined in cysteine (20).

Methyl Alkyl, Methyl Branched Alkyl, and Methyl Alkenyl Disulfides (Table 8.10) (30)

In the absence of conformational isomerism, only two bands are observed between 500 and 700 cm^{-1} for the CH_3-S stretch and S-S stretch of methyl disulfide [Figure 8.4(A)]. Depolarization ratios for C-S bands generally are significantly higher than those for corresponding S-S bands. The spectrum of ethyl methyl disulfide [Figure 8.4(B)] is more complex because of bands arising from different conformations about the C-S bond. These are evident from the two pairs of bands at 672, 643 cm^{-1} (C-S stretch) and 526, 511 cm^{-1} (S-S stretch). Since the 526 and 672 cm^{-1} bands decrease in strength relative to the lower frequency components on cooling, they represent the higher-energy, probably *gauche*, form. Similar to ethyl methyl disulfide, the higher frequency C-S and S-S rotamer bands in methyl propyl disulfide [Figure 4.8(C)] do not persist in the solid. The shoulder at 705 cm^{-1} may originate from a conformer(s) about the CC-CS bond. The scattering pattern of methyl propyl disulfide appears to be characteristic of methyl butyl and higher disulfides. Methyl isopropyl [Figure 8.4(D)] and methyl isobutyl analogs expectedly present different profiles between 500 and 700 cm^{-1}.

The spectrum of a mixture of (Z)- and (E)-propenyl disulfide, isolated from onion oil, is instructive (Figure 8.5).

TABLE 8.10. Raman Spectral Data for Some Methyl Alkyl, Methyl Branched Alkyl, and Methyl Alkenyl Disulfides (30)

Disulfide*	Structure	ν_{C-S} (cm^{-1})	ρ	ν_{S-S} (cm^{-1})	ρ
Methyl *1SS1*	C-S-S-C	691[b]	0.15	505[a]	0.07
Methyl propyl *3SS1*	C-C-C-S-S-C	705 sh 692 650[b] 631	- 0.15 - 0.13	526 511[a]	0.10 0.06
Isopropyl methyl *1YSS1*	C-C(C)-S-S-C	691[a] 623 596	0.16 0.12 0.14	525[a] 510	0.08 0.03
tert-Butyl *1XSS1*	C-C(C)(C)-S-S-C	688 571[a]	0.19 0.12	526[b]	0.08
(Z)-and (E)-Methyl propenyl *2U1SS1*	C-S-S-C=C-C	694[b]	0.13	515 497[a]	0.09 0.06

*Wiswesser Line Notation in italics
[a]Strongest spectral band
[b]Second strongest spectral band

The 1623 and 1612 cm^{-1} bands (ρ ∿0.1 and ∿0.06, respectively) originate in the (E) and (Z) ethylenic groups. The band at 1377 cm^{-1} arises from the ethylene methyl moiety and the strong 694 cm^{-1} band (ρ = 0.13) is attributed to the CH_3-S stretching vibration. Scattering at 517 and 497 cm^{-1} (ρ = 0.09 and 0.06, respectively) are ascribed to the *gauche* and *trans* conformers; only the latter is observed in the Raman spectrum obtained on the sample in solid state. The only candidate for the S-CH= stretch appears at 555 cm^{-1} (ρ = 0.2), a frequency lower than expected for this vibrational mode. The Raman spectrum of methyl propenyl sulfide also shows a weak band at 550 cm^{-1} (ρ = 0.19); no scattering is observed between 575 and 700 cm^{-1}.

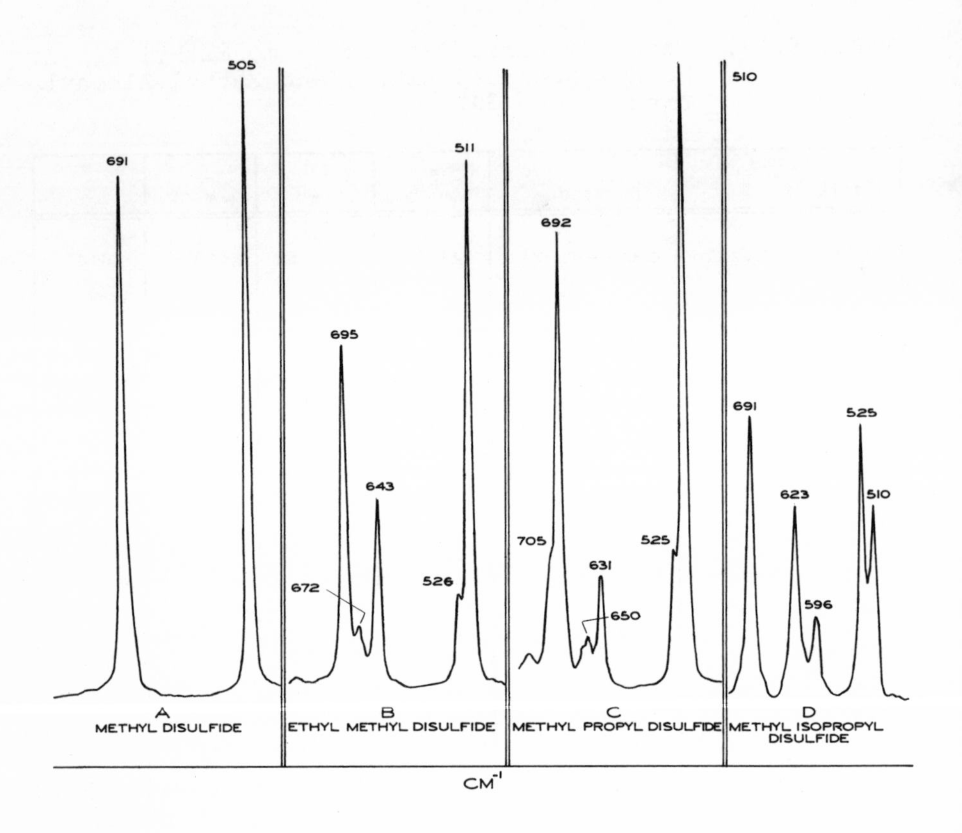

Figure 8.4. Raman spectra of some methyl alkyl and branched alkyl disulfides (30).

Alkyl Disulfides Other Than Methyl

Straight Chain Disulfides. A very simple profile is observed for ethyl disulfide, [Figure 8.6(A)] and the original data of Scott et al. (11) was confirmed recently (30). A large increase in ρ values for the C-S stretches occurs for the two conformers compared to those of ethyl sulfide (Tables 8.6 and 8.11). The more stable TT rotamer is evidenced by the band set at 507 and 640 cm^{-1}, and the higher-energy TG by bands at 525 and 668 cm^{-1}. Substitution of a propyl for an ethyl group of ethyl disulfide introduces additional, spectroscopically active rotational isomers [Figure 8.6(B)].

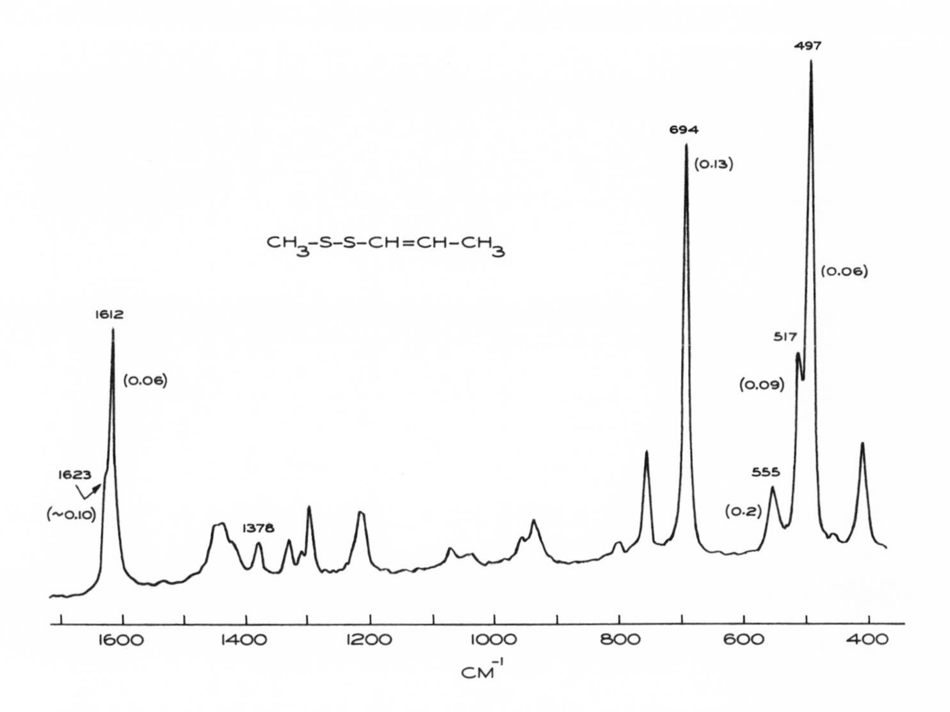

Figure 8.5. Raman spectrum of a mixture of (Z)- and (E)-methyl propenyl sulfide isolated from onion oil (30).

The new band appearing at 701 cm^{-1} (ρ = 0.3) probably represents one or more of these rotamers. Assigning this band to conformers about the CC-CS bond is in agreement with preceding comments pertaining to the propyl group in sulfides. The weaker, higher frequency bands of the pairs situated at 504 cm^{-1} (ρ = 0.09), 518 cm^{-1} (0.11), 637 cm^{-1} (0.22), and 666 cm^{-1} (0.18) are ascribed, as above, to the higher-energy rotamers with reference to hindered rotation of the alkyl groups about the S-S and C-S bonds, respectively. The band series for propyl [Figure 8.6(C)], butyl, butyl ethyl, butyl propyl, and pentyl disulfudes between 500 and 705 cm^{-1} are quite similar and closely resemble that of ethyl propyl disulfide (Table 8.11). It appears that this characteristic profile, including depolarization values, can be employed to identify these and higher n-alkyl disulfides (30).

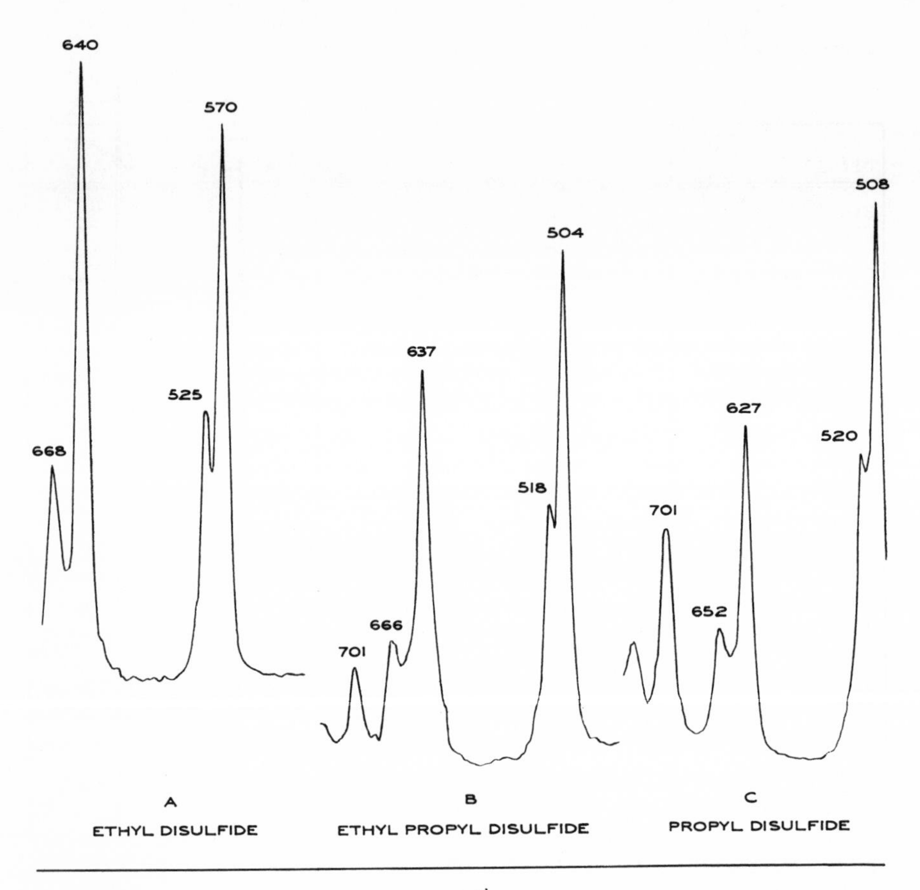

Figure 8.6. Raman profiles for some alkyl disulfides (30).

Cooling experiments serve to confirm the identifications. For example, spectra recorded on cooled (-110° C) butyl and pentyl disulfides show an appreciable decrease in the intensity of the weaker, higher frequency components of the C-S and S-S stretching vibration pairs; good evidence for attributing the bands to the less stable (*gauche*) rotamers. Frequency differences between the C-S and S-S conformer bands fall in the narrow ranges 25-29 and 12-14 cm^{-1}, respectively.

TABLE 8.11. Raman Spectral Data for Some Alkyl Disulfides (Other Than Methyl) (30)

Disulfide*	Structure	ν_{C-S} (cm^{-1})	ρ	ν_{S-S} (cm^{-1})	ρ
Ethyl *2SS2*	C-C-S-S-C-C	668 640[a]	0.20 0.22	525 507[b]	0.09 0.07
Ethyl propyl *3SS2*	C-C-S-S-C-C-C	701 666 637[b]	0.3 0.18 0.22	518 504[a]	0.11 0.09
Propyl *3SS3*	C-C-C-S-S-C-C-C	701 652 627[b]	0.22 ∿0.20 0.20	520 508[a]	∿0.1 0.08
Butyl *4SS4*	$n\text{-}C_4\text{-}S\text{-}S\text{-}C_4^n$	700 655 630[c]	0.28 0.15 0.20	520[b] 508[a]	∿0.10 0.10
Butyl ethyl *4SS2*	$n\text{-}C_4\text{-}S\text{-}S\text{-}C\text{-}C$	702 668 639[b]	∿0.2 - 0.19	522 508[a]	0.10 0.07
Butyl propyl *4SS3*	$n\text{-}C_4\text{-}S\text{-}S\text{-}C\text{-}C\text{-}C$	705 660 631[c]	∿0.2 0.16 0.22	525[b] 512[a]	0.11 0.10
Pentyl *5SS5*	$n\text{-}C_5\text{-}S\text{-}S\text{-}C_5^n$	700 658 630[c]	0.30 0.16 0.19	522 508	0.09 0.08
Hexyl *6SS6*	$n\text{-}C_6\text{-}S\text{-}S\text{-}C_6^n$	702 658 632	0.25 ∿0.1 0.10	522 510	0.10 0.09

*Wiswesser Line Notation in italics

[a]Strongest spectral band below 1400 cm^{-1}.

[b]Second strongest spectral band below 1400 cm^{-1}.

[c]Third strongest spectral band below 1400 cm^{-1}.

Branched Chain Alkyl and Alkenyl Disulfides. The 475-725 cm^{-1} scattering regions of isopropyl, *tert*-butyl, and 1,2-dimethylpropyl disulfides differ considerably from the straight chain analogs (Table 8.12) (30), but when branched methyl groups are not situated adjacent to the sulfur atoms the spectral profiles resemble the characteristic n-alkyl compounds. Note that here again the allyl C-S stretch appears at abnormally high frequencies.

tert-Butyl Disulfide. Based on an IR study (25), a

TABLE 8.12. Raman Spectral Data for Some Branched Chain Alkyl and Alkenyl Disulfides (30)

Disulfide*	Structure	ν_{C-S} (cm^{-1})	ρ	ν_{C-S} (cm^{-1})	ρ
Isopropyl *1Y&SSY*	C–C(C)–S–S–C(C)–C	625[a] 598	0.14 0.20	542 528	0.11 0.09
sec-Butyl *2Y&SSY2*	C–C–C(C)–S–S–C(C)–C–C	678[a] 650 635 628	0.16 - - 0.14	542 524[b] ~510 sh -	0.20 0.08 - -
1,2-Dimethyl-propyl *1Y&Y&SSY&Y*	C–C(C)–C(C)–S–S–C(C)–C(C)–C	672[a] 660 sh 648	0.11 - 0.13	520 ~508 sh -	0.09 - -
Isobutyl *1Y&1SS1Y*	C–C(C)–C–S–S–C–C(C)–C	738 702[a] 660 wk	0.16 0.19 -	520 508[c] -	0.13 0.1 -
Ethyl isobutyl *2SSY*	C–C(C)–S–S–C–C	705 668 639	0.20 0.18 0.23	522 508	0.11 0.10
tert-Butyl ethyl *2SSX*	C–C(C)(C)–S–S–C–C	663 638 570[a]	0.19 0.21 0.14	540 sh 525 -	- 0.08 -
Isobutenyl *1Y&U1SS1UY*	C–C(C)=C–S–S–C=C(C)–C	645 sh 642[a] -	- 0.15 -	525 514 498[b]	- - -
Allyl *1U2SS2U1*	C=C–C–S–S–C–C=C	750 720[a]	0.26 0.26	510[b] -	0.11 -
tert-Butyl *1X&&SSX*	C–C(C)(C)–S–S–C(C)(C)–C	572[a]	0.09	545[c]	0.07

"strong" band at 568 cm^{-1} was assigned to the *tert*-butyl-S stretching vibration, but no candidate could be found for the S-S stretch. Actually, the C-S band is quite weak in the IR (20), but it is strong in the Raman effect and falls in the range previously discussed for the thiol and sulfide. Furthermore, the depolarization value associated with this vibration (0.09) expectedly is approximately twice that of the similar mode in *tert*-butyl disulfide. The moderate intensity 545 cm^{-1} Raman band, present in the spectra of ethyl (570 cm^{-1}) and methyl (571 cm^{-1}) *tert*-butyl disulfides, is ascribed to the S-S stretching vibration (30). It is observed in the IR spectrum as a

TABLE 8.12. (Continued)

Disulfide*	Structure	ν_{C-S} (cm^{-1})	ρ	ν_{S-S} (cm^{-1})	ρ
Allyl propyl *3SS2U1*	C=C-C-S-S-C-C-C	750 718[b] 702 650 628	- 0.15 0.23 - 0.16	522 sh 508 - - -	- 0.07 - - -
Cystine *QVYZ1S 2*	$(-S-C-C(NH_2)-COOH)_2$	678	-	498[a]	-
Bis (dimethyl-thiocarbamoyl) (Tetramethyl-thiuram) *1N1&YUS&S 2*	$(C-N(C)-C(=S)-S-)_2$	-	-	562	-

*Wiswesser Line Notation in italics

[a]Strongest spectral band below 2800 cm^{-1}.

[b]Second strongest spectral band below 2800 cm^{-1}.

[c]Third strongest spectral band below 2800 cm^{-1}.

[wk]Weak band.

weak shoulder at 543 cm^{-1}.

Aryl and Aralkyl Disulfides. The S-S stretching frequencies for phenyl (25) and phenalkyl disulfides are not appreciably different (Table 8.13) (30), but thienyl and furyl disulfides display abnormally low S-S frequencies (about 465 cm^{-1}).

8.4.2 Cyclic Disulfides

Based on the Raman spectra of several cyclic disulfides (Table 8.14) (20) the S-S stretching band appears to lie between 490 and 550 cm^{-1}. Changes in the dihedral angle are not accompanied by changes in the S-S/C-S band intensity ratios (see p.233). For example, compare thioctic acid (θ = 27°) (31) and 1,2-dithiepane (θ = 90°) (32).

TABLE 8.13. Raman Spectral Data for Some Aryl and Aralkyl Disulfides (30)

Disulfide*	Structure	ν_{C-S} (cm^{-1})	ρ	Comments	ν_{S-S} (cm^{-1})	ρ
Phenyl *RSSR*	φ–S–S–φ	695[b]	0.24	"X-sensitive"	523[a]	0.09
Benzyl *R1SS1R*	φ–C–S–S–C–φ	659[a]	0.12	-	505[b]	0.13
Benzyl Methyl *1SS1R*	φ–C–S–S–C	692 655	0.17 0.18	CH-S stretch C-S stretch	508[b]	0.10
2,2'-Dithienyl *T5SJ BS 2*	S–S	640	-	-	466[c]	-
5,5-Dimethyl- 2,2'-dithienyl *T5SJ B ES 2*	S–S	675 (?)	-	Very weak	465[c]	-
Furfuryl *T5OJ B1S 2*	C–S–S–C	606[b]	0.16	-	520	0.11
2-Methyl-3-furyl *T5OJ B CS 2*	S–S	677	0.16	-	464	0.09

*Wiswesser Line Notation in italics

[a]Strongest band nelow 1400 cm^{-1}.

[b]2nd Strongest band below 1400 cm^{-1}.

[c]Strongest band below 1400 cm^{-1}.

Differences in dihedral angles of disulfides are reflected by changes in UV spectra (31): open chain, $\theta = 90°$, $\lambda_{max} = 250$ nm; 1,2-dithianes, $\theta = 60°$, $\lambda_{max} = 255$ nm; 1,2-dithiolanes, $\theta = 27°$, $\lambda_{max} = 330$ nm. Consequently, the six- and five-membered ring 1,2-disulfides can be distinguished by UV.

Lenthionine, an odorous compound found in the Shiitake mushroom, contains both di- and trisulfide groups. The former appears at 502 cm^{-1} ($\rho = 0.05$) and the latter at 482 ($\rho = 0.05$). The C-S stretch is observed at 647 cm^{-1} ($\rho = 0.2$) (33).

8.4.3 Dependence of Intensity Ratios of S-S/C-S Stretching Bands on Bond and Dihedral Angles

Raman spectroscopy can provide direct evidence concerning the presence and number of disulfide linkages in proteins, and also is useful in studying the geometry of the CSSC group. Attention has been called to the weakness of the C-S stretching band of the CSSC group in the Raman spectrum of native lysozyme compared to the intense S-S stretch at 509 cm^{-1} (34). A superposition of the Raman spectra of lysozyme's constituent amino acids demonstrated that the C-S stretching band was quite strong by comparison with that of lysozyme. From this behavior, it was inferred that the conformations of the C-S-S-C cross-links in lysozyme are different from those in the monomeric analogs in solution. These widely divergent intensity differences may be related to the C-S-S bond angle (34). For example, the S-S/C-S stretching frequency intensity ratio is 10/1 for crystalline cystine (114.5° bond angle), while it is about 0.5/1 for cystine·HBr (104°). In an aqueous acid solution of cystine this ratio is 1.5/1.

One knows that the disulfide bonds of insulin remain intact in the denatured state from the fact that the S-S stretching frequencies centered at 516 cm^{-1} do not weaken or disappear (35). The C-S stretching vibrations of the disulfide moiety in native insulin appear at 668 and 680 cm^{-1}. Striking spectral changes occur on denaturation in the 670 cm^{-1} region and alterations in band shape and peak intensity of the 516 cm^{-1} band strongly suggest that the dihedral angles and the CSS bond angles have changed appreciably.

TABLE 8.14. Raman Spectral Data for Some Cyclic Disulfides (20)

Chemical Name*	Structure	ν_{C-S}			ν_{S-S}		
		cm^{-1}	ρ	Int.	cm^{-1}	ρ	Rel. Int.
3-Methyl-1,2-dithiolane *T5SSTJ C*		678 649	- -	6 4	518 511	- -	8 10
1,2-Dithiolane-3-valeric acid (Thioctic acid) *T5SSTJ C4VQ*	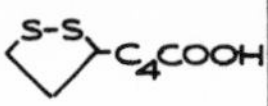	680 635	- -	4 3	509	-	10
o-Dithiane *T5SSTJ*		560	0.10	8	510	0.10	10
4-Chloro-5-morpholino-3H-1,2-dithiole-$\Delta^{3,N}$ carbamic acid, methyl ester *T6N DCTJ A- ET5SSYJ CUNVO1 DG*	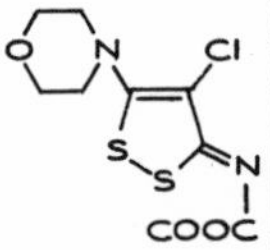	-	-	-	490	-	10
1,2-Dithiepane *T7SSTJ*		638 630	0.15 0.13	5 4	513	0.06	10
1,2,4-Trithiolane *T5SS DSTJ*	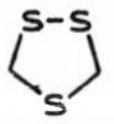	668	0.05	6	509	0.11	10
3,5-Dimethyl-1,2,4-trithiolane *T5SS DSTJ C E*	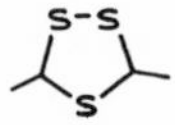	610	0.08	6	517	0.09	
1,2,5-Trithiepane *T7SS ESTJ*		608	0.04	9	510	0.05	10
5-Methyl-1,2,5-trithiepane iodide *T7SS ESTJ E &I*		608	-	4	505	-	10
6,7,13,14-Tetrathiadispiro [4.2.4.2]tetradecane *T6SSXSSXJ C-& F-&/ AL5XTJ 2*	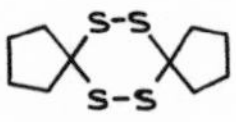	-	-	-	541	-	9
7,8,15,16-Tetrathiadispiro-[5.2.5.2]hexadecane *T6SSXSSXJ C-& F-&/ AL6XTJ 2*	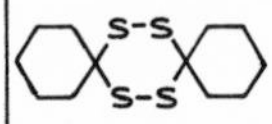	-	-	-	547	-	10
1,2,3,5,6-Pentathiepane (Lenthionine) *T7SSS ESSTJ*	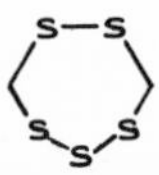	647	∼0.2	2	502 482	0.05 0.05	7 10

*Wiswesser Line Notation in italics.

The relative intensities of S-S and C-S bands may be diagnostic of alkyl substitution on carbon atoms contiguous to the sulfurs (30). Values for some symmetrical and unsymmetrical straight chain disulfides range between 1 and 2.0 (Table 8.15) (30).

TABLE 8.15. Intensity Ratios of S-S and C-S Stretching Vibrations (30)

Disulfide	S-S/C-S
Methyl	1.2
Ethyl	0.9
Ethyl propyl	1.2
Propyl	1.6
Butyl	1.9
Butyl ethyl	1.0
Butyl propyl	1.9
Pentyl	2.0
Hexyl	1.9
Ethyl methyl	1.5
Methyl propyl	1.4
Methyl propenyl	1.2
Isopropyl methyl	1.0
Isopropyl	0.35
Isobutyl	0.5
1,2-Dimethylpropyl	0.32
Ethyl isobutyl	1.0
tert-Butyl methyl	1.1
tert-Butyl ethyl	1.5
tert-Butyl	0.38
Allyl	1.3
Allyl propyl	1.8
Cystine	10
Cystine HCl	1.5

An appreciable decrease occurs when methyl groups are situated on the 1-carbon atoms, that is, isopropyl, *tert*-butyl, and 1,2-dimethylpropyl disulfides (0.32 - 0.38). When only one of these carbons is substituted, values lie near 1.3. An intermediate ratio of 0.5 is found for isobutyl disulfide, where the methyls are located on the second carbons removed from the sulfur atoms. It is possible that a steric effect is operating which increases the C-S-S-C dihedral angle, resulting in a decrease in the intensities of the S-S stretching bands relative to the C-S stretches. However, severely decreasing the normal 90° dihedral angle by incorporation of the disulfide linkage in cyclic systems does not change the ratio of about 1.5. It has been suggested that the absorption band of alkyl disulfides near 250 nm is related to the electron cloud interaction responsible for the energetically preferred conformations (36) (approximately 90° dihedral angle). The suppression of this absorption observed in the case of *tert*-butyl disulfide would then be expected as a result of steric hindrance, forcing the angle to exceed 90°. Scale models indicate that the dihedral angle is near 100°.

8.5 TRISULFIDES

8.5.1 Alkyl Trisulfides

Owing to the fact that unsymmetrical alkyl trisulfides readily rearrange to the symmetrical isomers, no spectral data have been compiled on the former class.

Methyl Trisulfide

As anticipated, a simple profile is displayed by this trisulfide [Figure 8.7(A)]. The CH_3-S stretch has essentially the same spectral location and ρ value (697 cm^{-1}, 0.16) as in methyl disulfide. The S-S-S stretch appears as a symmetrical band at 487 cm^{-1} ($\rho = 0.06$). However, when a polarizing analyzer is oriented perpendicularly, a new band appears at 482 cm^{-1} whose intensity is approximately equal to the one at 487 cm^{-1}. The former may arise from the asymmetric S-S-S stretch, while the latter is attributed to the symmetric stretch (20).

Ethyl Trisulfide

Two C-S conformers manifest themselves at 670 cm^{-1} ($\rho = 0.20$) and 642 cm^{-1} (0.20), the former originating from the less stable rotamer, since its intensity is substantially reduced on cooling (20) [Figure 8.7(B)]. Presence of the S-S-S rotamers is evidenced by a shoulder at 500 cm^{-1} ($\rho \sim 0.1$), which nearly disappears on cooling, and an intense 487 cm^{-1} ($\rho = 0.08$).

Propyl Trisulfide

The spectrum [Figure 8.7(C)] resembles that of ethyl trisulfide with respect to the C-S rotamers: 650 cm^{-1} (ρ = 0.20), 630 cm^{-1} (ρ = 0.20), and the S-S-S rotamers: 498 cm^{-1} (ρ = 0.09), 485 cm^{-1} (ρ = 0.09). In the solid, the high frequency components of each pair are absent from the spectrum (20). Similar to previously discussed sulfides and disulfides containing a propyl group, a band is present at 704 cm^{-1} (ρ ∿0.2), correlatable with the rotamer(s) about the C-CS bond.

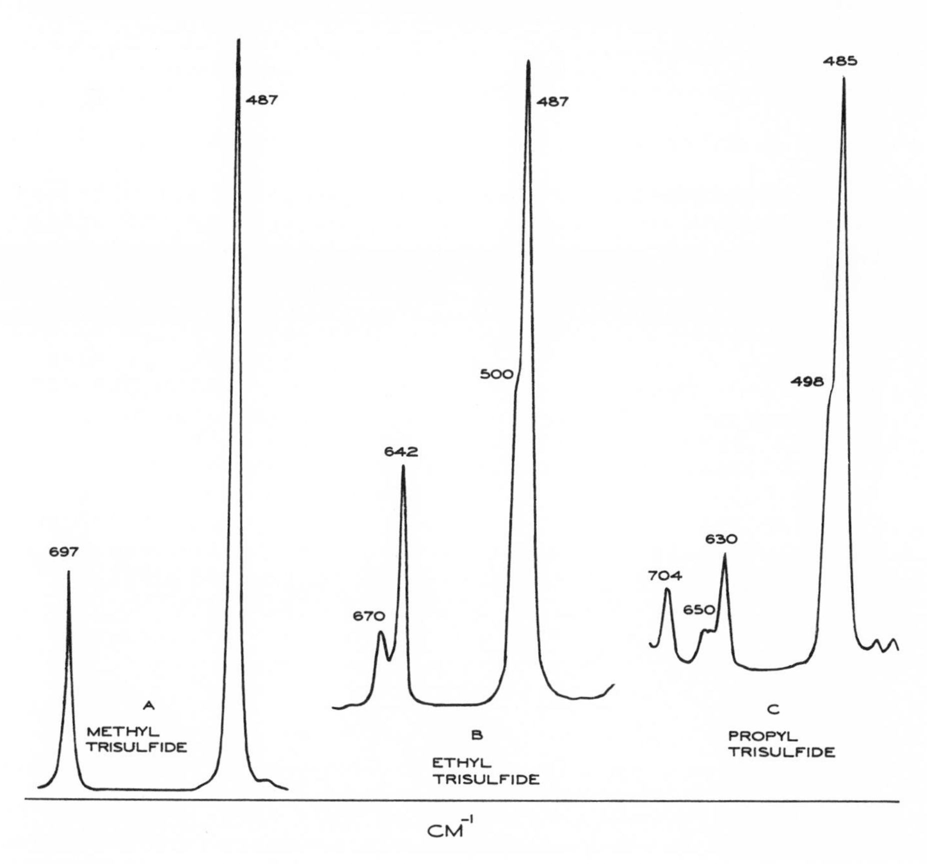

Figure 8.7. Raman spectra of some alkyl trisulfides (20).

8.5.2 Cyclic Trisulfides

The frequencies of the C-S, S-S, and S-S-S stretching modes in lenthionine (20) (Table 8.14) agree well with values for acyclic di- and trisulfides.

8.6 ACYCLIC POLYSULFIDES

Simple tetra- and pentasulfides display S-S stretching bands of lower frequency than di- and trisulfides (37) (Table 8.16).

8.7 VULCANIZED RUBBERS

Raman spectroscopy offers considerable potential for the study of vulcanized rubber. The major structural changes involving C-S and S-S groups which are believed to occur on accelerated or unaccelerated sulfur vulcanizations are summarized in Table 8.17 (38). It appears feasible to detect, and possibly measure, the amount of mono-, di-, and polysulfide cross-links without chemically modifying the network. Good quality Raman spectra have been obtained from vulcanizates prepared from (Z)-1,4-polybutadiene and significant differences may be observed in the spectra of extracted vulcanizates prepared in different ways (39).

The Raman spectrum from 400 to 900 cm^{-1} of the extracted vulcanizate shows new Raman bands at 708, 635, and 505 cm^{-1} (40). These emissions lie close to those of 2,5-dihydrothiophene (711, 641, 506 cm^{-1}), which suggests the presence of I in the vulcanizate. Confirmation of this cyclic structure would have major significance with regard to the mechanism of accelerated sulfur vulcanization.

```
  CH2   CH=CH   CH2        CH2
 /   \ /     \ /   \      /   \
      CH      CH    CH=CH
        \    /
          S
```

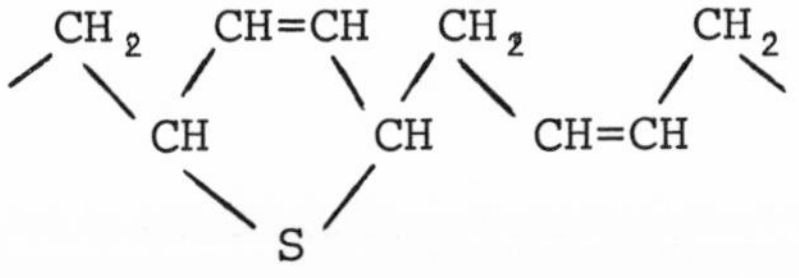

I

TABLE 8.16. Raman Spectral Data for Some Polysulfides (37)

Chemical Name*	Structure	$\nu_{S\text{-}S}$ stretch (cm^{-1})	Rel. Int.
Methyl Disulfide *1SS1*	$C\text{-}S_2\text{-}C$	505	10
Methyl Trisulfide *1SSS1*	$C\text{-}S_3\text{-}C$	487	10
Methyl Tetrasulfide *1SSSS1*	$C\text{-}S_4\text{-}C$	487 441	(very strong) (very strong)
Ethyl Disulfide *2SS2*	$C_2\text{-}S_2\text{-}C_2$	525 507	4 10
Ethyl Trisulfide *2SSS2*	$C_2\text{-}S_3\text{-}C_2$	487	10
Ethyl Tetrasulfide *2SSSS2*	$C_2\text{-}S_4\text{-}C_2$	486 438	(very strong) (strong)
Ethyl Pentasulfide *2SSSSS2*	$C_2\text{-}S_5\text{-}C_2$	488 456 435	(very strong) (strong) (strong)

*Wiswesser Line Notation in italics

In spite of the initial encouraging results, many problems exist in studying rubber systems by Raman spectroscopy. Difficulties are associated with sample fluorescence and degradation under the high-energy sources necessary to obtain spectra, especially when weak bands of importance are present and when quantitative measurements of band intensities are made.

TABLE 8.17. Some Structural Features of Sulfur Vulcanized Rubbers (38)

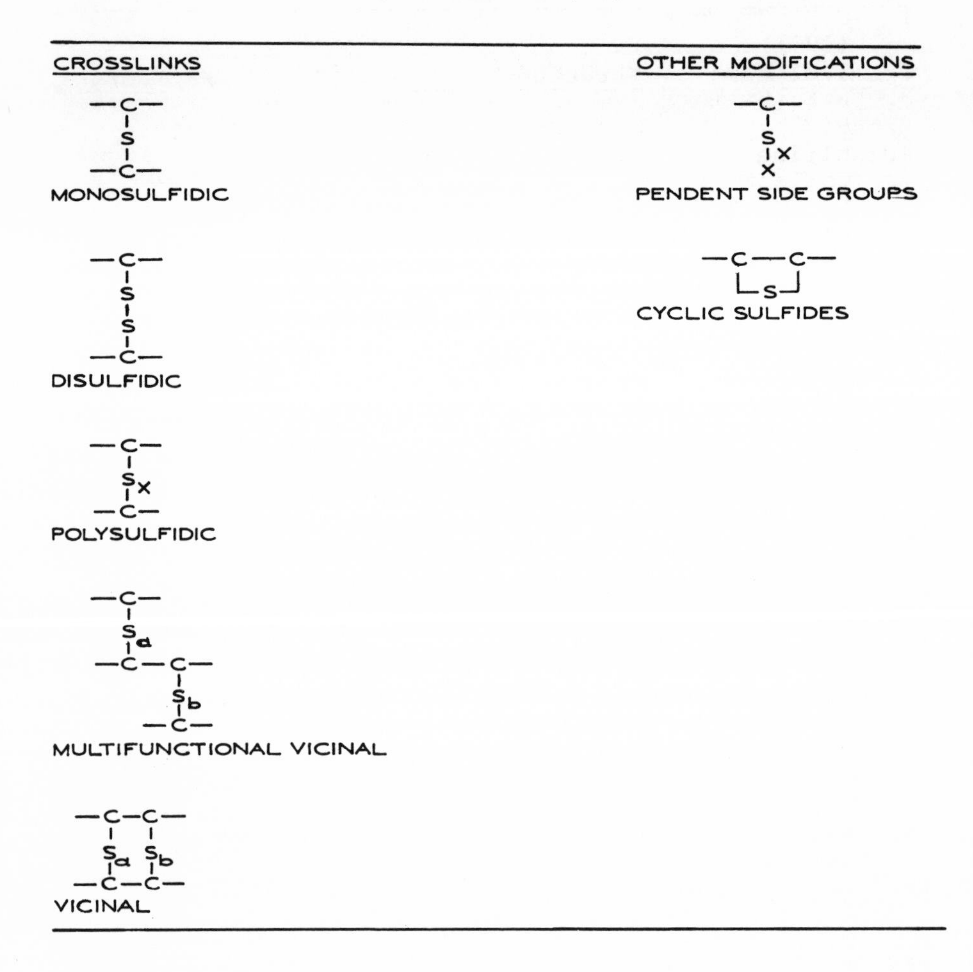

8.8 THIOACETATES

Ethyl through heptyl thioacetates exhibit intense Raman bands at 627±2 cm^{-1} (ρ = 0.04-0.06) attributed to the C-S stretching vibration (20).

REFERENCES

1. G.D. Gal'pern, Int. J. Sulfur Chem. B6, (2), 115 (1971).
2. Z.M. Bacq, Int. J. Sulfur Chem. B6, (2) 93 (1971).
3. R.F. Steiner, *The Chemical Foundations of Molecular Biology* (Van Nostrand, Princeton, N.J., 1965).
4. K. Jost, V. Debabov, H. Nesbavda, and J. Rudinger, Collect. Czech. Chem. Comm. 29, 419 (1964).
5. L. Young and G.A. Maw, *The Metabolism of Sulphur Compounds* (Wiley, New York, 1958).
6. S. Schwimmer and M. Friedman, The Flavour Industry 3, 137 (1972).
7. S. Venkateswaran, Ind. J. Physics 6, 51 (1931).
8. A.S. Ganesan, Phil. Mag. 15, 51 (1933).
9. I.F. Trotter and W.H. Thompson, J. Chem. Soc. 481 (1946).
10. N. Sheppard, Trans. Faraday Soc. 46, 429 (1950).
11. D.W. Scott, H.L. Finke, J.P. McCullough, M.E. Gross, R.E. Pennington, and G. Waddington, J. Am. Chem. Soc. 74, 2478 (1951).
12. F. Fehér and H.J. Berthold, Chem. Ber. 88, 1634 (1955).
13. D.W. Scott and J.P. McCullough, J. Am. Chem. Soc. 80, 3554 (1958).
14. E.M. Popov and G.I. Kagan, Optics and Spectroscopy 11, 394 (1961).
15. P.G. Puranik and V. Kumar, Current Sci. (India) 31, 179 (1962).
16. J.P. Perchard, M.T. Forel, and M.L. Josien, J. Chim. Phys. 61, 645 (1964).
17. H. Gerding and J.W. Ypenburg, Recueil 85, 616 (1966).
18. R.C. Laughlin and W. Yellin, J. Am. Chem. Soc. 89, 2435 (1967).
19. L.J. Bellamy, *The Infra-red Spectra of Complex Molecules* (Wiley, New York, 1956).
20. S.K. Freeman (unpublished work).
21. T. Torgrimsen and P. Klaeboe, Acta Chemica Scan. 24, 1139 (1970).
22. S.K. Freeman and D.W. Mayo, Appl. Spec. 27, 286 (1973).
23. M. Hayashi, T. Shimanouchi, and S. Mizushima, J. Chem. Phys. 26, 608 (1957).

24. D.W. Scott, H.L. Finke, W.N. Hubbard, J.P. McCullough, G.D. Oliver, M.E. Gross, C. Katz, K.D. Williamson, G. Waddington, and H.M. Huffman, J. Am. Chem. Soc. 74, 4656 (1952).
25. K.G. Allum, J.A. Creighton, J.H.S. Green, G.J. Minkoff, and L.J.S. Pierce, Spectrochim. Acta 24A, 927 (1968).
26. S.K. Freeman and D.W. Mayo, Paper presented at the Pittsburgh Symposium, Cleveland, Ohio, March 1971.
27. P.A. Akishin, N.G. Rambidi, I.N. Tit-skvortosva, and Yu. K. Yur'ew, Sbornik Trudov Mezhvuz. Soveshcha. po Khim. Nefti, Moscow 1956 146-62 (Publ. 1960).
28. R.R. Fraser, G. Boussard, J.K. Saunders, J.B. Lambert, and C.E. Mixan, J. Am. Chem. Soc. 95, 3822 (1971).
29. S.G. Frankiss, J. Mol. Str. 3, 89 (1969).
30. S.K. Freeman and D.W. Mayo, Paper presented at the Eastern Analytical Symposium, Atlantic City, N.J., November 1972.
31. R. Raiman, S. Safe, and A. Taylor, Quart. Rev. Chem. Soc. 24, 208 (1970).
32. G. Claeson, G. Androes, and M. Calvin, J. Am. Chem. Soc. 83, 4357 (1961).
33. S.K. Freeman, J. Agr. Food Chem. 21, 521 (1973).
34. R.C. Lord and N.J. Yu, Mol. Biol. 50, 509 (1970); 51, 203 (1970).
35. N.-T. Yu and C.S. Liu, J. Am. Chem. Soc. 94, 3250 (1972).
36. H.P. Koch, J. Chem. Soc. 394 (1949).
37. F. Fehér, G. Krause, and K. Vogelbruch, Chem. Ber. 90, 1570 (1957).
38. L. Bateman (Ed.), *The Chemistry and Physics of Rubber-like Substances* (MacLauren, London, 1963), Chap. 15.
39. J.L. Koenig, M.M. Coleman, J.R. Shelton, and P.H. Starmer, Rubber Chem. and Tech. 44, 71 (1971).
40. M.M. Coleman, J.R. Shelton, and J.L. Koenig, Rubber Chem. and Tech. 45, 173 (1972).

Chapter 9

SYNTHETIC POLYMERS

9.1 INTRODUCTION

The first Raman spectrum of a polymer (polystyrene) was obtained in 1932 (1), but nearly twenty years elapsed before the second polymer spectrum--poly (methyl methacrylate)--was reported (2). In fact, less than a dozen polymer molecules had been studied with Raman spectroscopy prior to the introduction of the laser. The potential value of Raman spectroscopy in this field was fully appreciated, but the dearth of publications was caused by the fact that most polymer samples do not possess those characteristics necessary for pre-laser Raman recording, that is, transparency and nonfluorescence. The first high-quality spectrum was generated in 1967 (3), shortly after the introduction of a laser source, and the past few years have seen numerous publications describing the applications of Raman spectroscopy in the synthetic polymer field, spearheaded by Koenig's group at Case Western Reserve University (4). Raman spectroscopy is potentially one of the most powerful techniques for polymer characterizations.

Functional groups giving rise to strong Raman bands can be used for structure determination and for quantitative analysis. The type of ethylenic unsaturation, that is, (Z), (E), or vinyl can be ascertained in butadiene (5) and isoprene (6) rubbers (see p.135). End group analysis by means of the vinyl C=C stretching band is a potential application of Raman spectroscopy, and ethylenic unsaturation can be quantitated for cross-linked systems. The intense scattering of thiols, sulfides, and disulfides makes possible the study of polymers containing sulfur (7) (see p.202). Aromatic rings incorporated in synthetic polymers should serve as the basis for Raman investigations in these systems.

9.2 QUANTITATIVE ASPECTS

Differing from IR spectroscopy, where the intensity of an absorption band is logarithmically proportional to sample concentration, the intensity of a Raman band is linear with concentration. For solution work, a known amount of "internal standard" can be added to assay the relative amount of a particular material in solution. The number of scattering units must be known for quantitative analysis of a solid, but this value cannot be determined by thickness measurements because Raman scattering occurs from the surface of a sample. In many instances it is convenient to select a Raman band arising from the sample itself as an internal standard band.

9.2.1 Vinyl Chloride in Vinyl Chloride--Vinylidene Copolymers

Quantitative measurements have been made on spectra generated from pellets of copolymers of vinyl chloride and vinylidene chloride (Saran wrap) employing the CH_2 asymmetric stretching frequency band (2926 cm^{-1}) common to both monomers as a scattering reference (Figures 9.1 and 9.2) (8). The band intensity at 2906 cm^{-1} is divided by the intensity at 2926 cm^{-1} (S_R). The 2906 cm^{-1} band is not present in the IR spectrum; the 1205 cm^{-1} IR absorption correlates with the vinyl chloride copolymer content up to 25% only. Figure 9.2 shows a linear relationship to 100% vinyl chloride using Raman spectroscopy. The composition can be measured to an accuracy of 2%. Linear relationships have been established between the copolymer, dyads, triads, and tetrads, and the intensity of certain Raman frequencies in the spectra of the copolymers.

9.2.2 Polystyrene in Nylons

The quantity of polystyrene in nylons has been determined by ascertaining the ratio of the intensities of the phenyl ring breathing band at 1003 cm^{-1} to the methylene deformation band at 1444 cm^{-1} (9). A linear relationship exists for polystyrene concentrations between 0 and 15%.

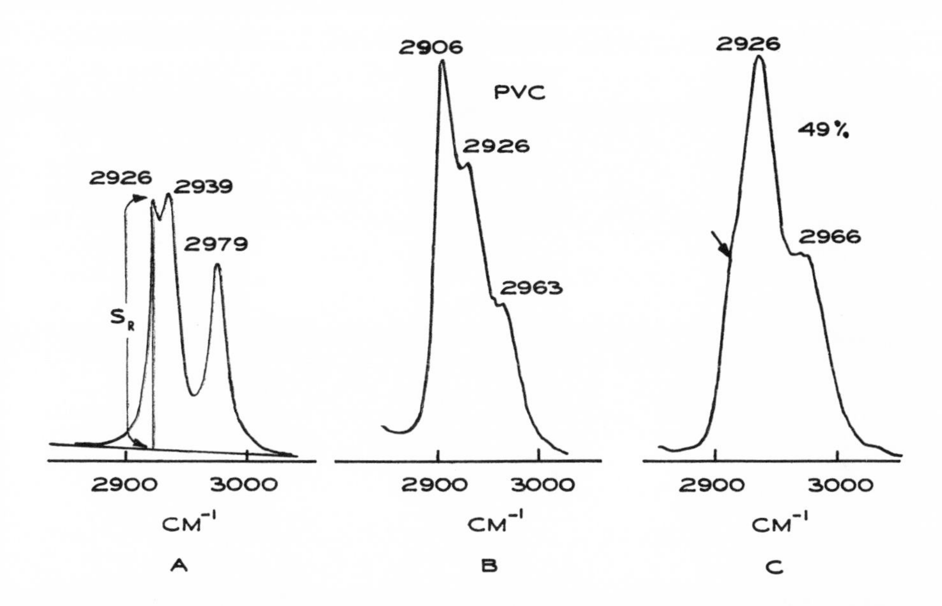

Figure 9.1. Raman spectra (carbon-hydrogen stretching region). (a) poly(vinylidene chloride)--S_R is the reference scattering intensity; (b) poly(vinyl chloride); (c) copolymer containing 49% vinyl chloride (8).

9.3 STEREOREGULARITY

9.3.1 Polyethylene Terephthalate (PET)

The relatively simple Raman spectrum of PET is quite sensitive to the state of crystallinity of the sample. This polymer has a wide range of crystallinity, from completely amorphous (0%) to 60%. Monofilaments and films, commercially prepared from the melt, are amorphous and crystallinity of a sample is created and enhanced by drawing and annealing. PET fibers important in the textile industry are "crease resistant," yet permanently creasable. The permanent pressing process is based on bending a fiber and heating to above its glass transition temperature. Raman spectroscopy provides a means for determining crystallinity because some of the spectral bands associated with the ester groups

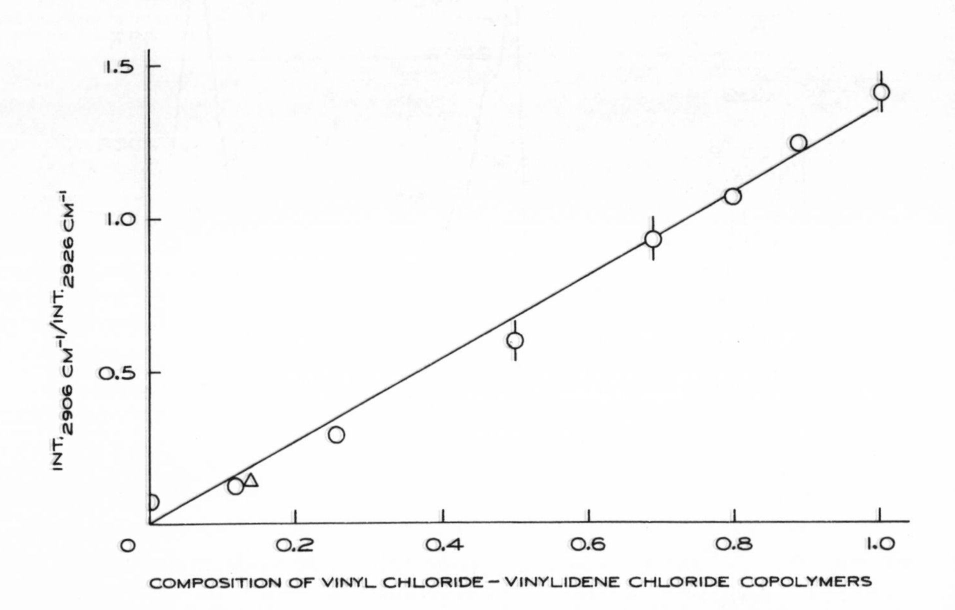

Figure 9.2. Standard curve for analysis of vinyl chloride-vinylidene chloride copolymers (9).

are very sensitive to crystallinity and/or orientation (10). Figure 9.3(A) shows the spectrum from fibers spun from the melt. When these are annealed above the glass transition temperature of PET, crystallization sets in, as shown by the sharpening of the Raman bands [Figure 9.3(B)]. Textile fibers are melt spun, drawn (to impart molecular orientation and therefore strength), and annealed (to impart good "wash and wear" characteristics). The Raman spectrum of such fibers [Figure 9.3(C)] indicates a higher degree of crystallization as opposed to 9.3(A) and 9.3(B), characterized by the appearance of a doublet near 1100 cm^{-1}, with the low frequency component stronger than the other.

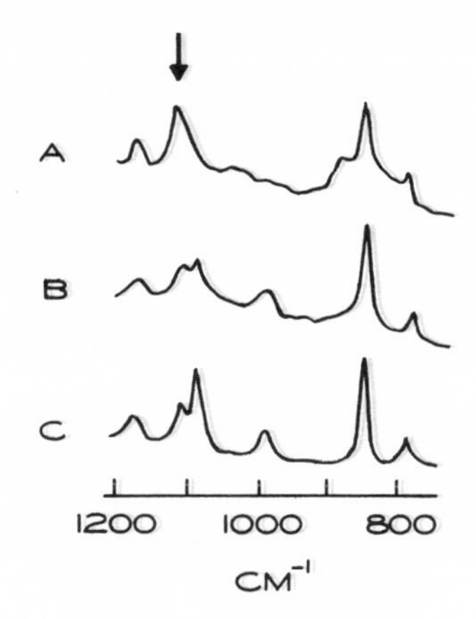

Figure 9.3. Raman spectra of polyethylene terephthalate (10).

The half-band width of the Raman C=O stretching vibration correlates with density in oriented and unoriented PET (11). In the crystalline polymer the successive carbonyl groups linked to the benzene ring adopt the *trans* conformation.

It appears that amorphous PET consists of a series of rotational states in which the carbonyls are rotated out of the plane of the benzene ring by varying degrees. This distribution of rotational isomers would tend to cause a broadening of the C=O stretching band, since each conformation would contribute its own C=O frequency. The structure is resonance stabilized as it tends toward planarity in the crystalline state, and a narrowing of the C=O band width occurs.

9.3.2 Polypropylene (PP)

Polypropylene samples are available in widely different tacticities and crystallinities (10). Changes in stereoregularity bring about changes in the Raman spectra. Samples with molecular weight around 10^5 give different spectra according to the degree of stereoregularity their molecules possess. In Figure 9.4(A), the spectrum of an oriented and annealed laminate of high crystallinity (~90%) isotactic PP contains a series of very sharp bands. An isotactic PP lump of lower crystallinity (~60%) gives a more diffuse spectrum as shown in Figure 9.4(B). The spectral detail is further decreased when an atactic PP lump of very low crystallinity is examined [Figure 9.4(C)].

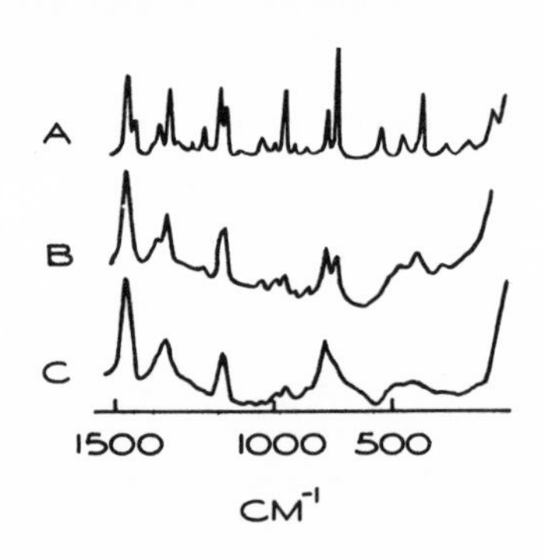

Figure 9.4. Raman spectra of polypropylene (10).

9.3.3 Ethylene-Propylene Rubber

The ethylene content of some ethylene-propylene rubbers has been found to vary from 25 to 80% (12). For copolymers containing more than 65% ethylene, a correlation exists between the degree of crystallinity determined by Raman spectroscopy and X-ray diffraction, and mechanical properties such as green strength.

9.3.4 Vinyl Polymers

The complementarity of Raman and IR spectroscopy permits distinguishing among the many varieties of possible stereoregular structures for monosubstituted vinyl polymers $(CH_2CHX-)_n$ (4). A polymer principally containing repeating units with identical configurations is termed "isotactic," while a polymer consisting mainly of exactly alternating configurational units is described as "syndiotactic." A chain showing no regular order of repeating unit configurations is termed "atatic." Not only can vinyl polymer structures be either isotactic or syndiotactic, but each of these stereoregular structures may adopt a planar or helical conformation. Alternatively, the polymer may possess an atactic form with a disordered conformation. All observed vibrational bands fall into particular groups, based on a classification into sets of vibrational modes and a subclassification according to their individual Raman polarization and IR dichroic behavior. The various possible structures and their idealized IR and Raman spectral characteristics are shown in Tables 9.1 and 9.2. Every structure considered has a unique set of spectroscopic properties. For example, the vibrational spectrum of a planar, isotactic vinyl polymer may be classified as [p, σ], [d, π], while a syndiotactic 3_1 helix is given by [d, σ], [p, 0], [0, π].

TABLE 9.1. Groups for Observed Vibrational Bands of Monosubstituted Vinyl Polymers (4)

[R, IR]	Raman	IR
[p, O]	Polarized	Inactive
[d, O]	Depolarized	Inactive
[O, π]	Inactive	Parallel dichroism*
[O, σ]	Inactive	Perpendicular dichroism*
[p, π]	Polarized	Parallel dichroism*
[p, σ]	Polarized	Perpendicular dichroism*
[d, π]	Depolarized	Parallel dichroism*
[d, σ]	Depolarized	Perpendicular dichroism*

*With respect to the helix axes

The value of this approach may be seen in the case of poly (vinyl chloride). Infrared (13,14) and X-ray (15) evidence indicate that this monosubstituted vinyl polymer exists in the extended syndiotactic form, but folded syndiotactic and threefold helical isotactic structures have been proposed. Referring to Table 9.2 and Figure 9.5 (16), it appears that the three structures can be differentiated by depolarization measurements. The [p, 0] class of spectroscopic activity is unique to the folded syndiotactic model for PVC and the spectral type [d, 0] is specific for the extended structure. For the helical isotactic model, the polarized Raman bands possess parallel IR dichroism, that is, [p, π]. The polarized bands in the Raman spectrum of ordinary PVC have counterpart IR absorptions, thereby ruling out the folded syndiotactic structure as an acceptable model, since this structure requires the polarized Raman bands to be unique.

Since the IR bands occurring at frequencies corresponding to the Raman polarized bands have perpendicular polarizations, that is [p, σ] type, the helical isotactic model is eliminated from consideration. The experimental observation of the [p, σ] band type confirms the X-ray and IR evidence for the extended syndiotactic chain. Additional support of this structure is gained from the observation of depolarized Raman bands, absent in the IR spectra, near the frequencies predicted from calculations; [p, σ].

9.4 CHAIN CONFORMATION IN THE SOLID STATE

Although Raman spectroscopy will not supplant X-ray diffraction for determining the conformation of a polymer chain in the solid state, it can be a helpful ancillary technique when fiber patterns cannot be obtained. Raman spectra of polymers usually are dominated by strong scattering arising from carbon-carbon skeletal vibrations. Conformational changes affect the frequencies of these backbone modes since they are highly coupled; any change in conformation will alter the coupling. Fortunately, differences between Raman and IR selection rules permit a determination for the planar, 2_1 and 3_1 helices. In general, the observed spectra will have modes exhibiting different frequencies because of the helix form (4). These modes are useful for characterizing helical conformations of a polymer in the solid state.

TABLE 9.2. Selection Rules for Monosubstituted Vinyl Polymers (4)

	Structure	Optical Activity								Example
	R:	p	p	d	d	p	d	O	O	
	IR:	π	σ	π	σ	O	O	π	σ	
	Center of Symmetry					√	√	√	√	Polyethylene, Polyethylene Sulfide
	Atactic	√	√	√	√					Polyvinyl Fluoride
Syndio-tactic	Helix $>3_1$				√	√	√	√		Polyethylene Oxide
Syndio-tactic	Helix 3_1				√	√		√		
Syndio-tactic	Helix 2_1			√	√	√				
Syndio-tactic	Planar		√	√	√		√			Polyvinyl Chloride
Isotactic	Helix 3_1	√			√		√			Polybutene
Isotactic	Helix 3_1	√			√					Polypropylene
Isotactic	Planar		√	√						

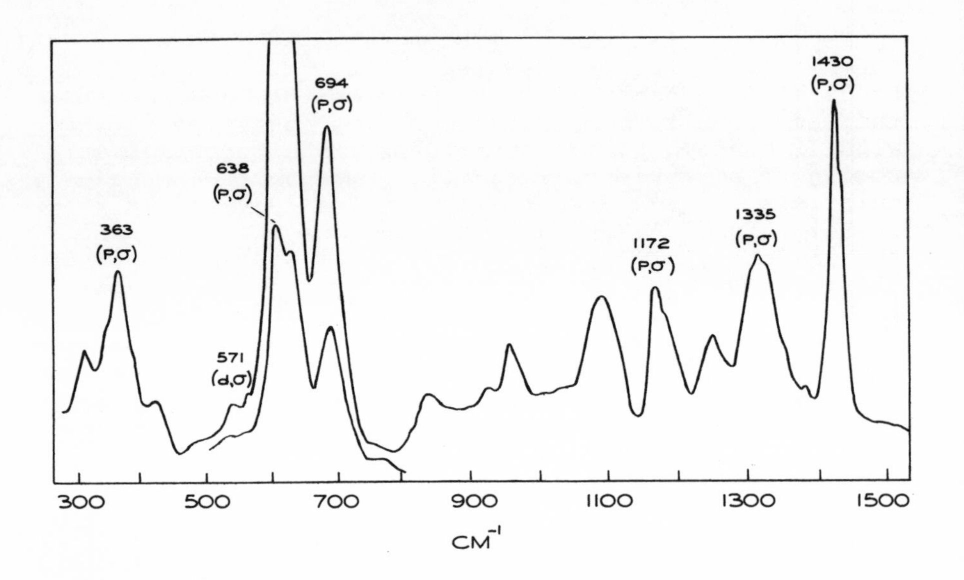

Figure 9.5. Raman spectrum of poly(vinyl chloride) (16).

9.4.1 Polytetrafluorethylene (PTFE)

Crystalline PTFE possesses a helical conformation with the number of CF_2 groups per repeating unit a function of temperature: below 19° the helix has 13 CF_2 groups in six turns of the helix (17), while above 19° there are 15 CF_2 groups in seven turns of the helix. Twenty-four vibrational modes have been calculated for crystalline PTFE, distributed among the following vibration types, or symmetry species (18):

$4A_1$: Raman active, polarized, strong Raman lines.
$3A_2$: IR active, parallel dichroism, Raman inactive.
$8E_1$: IR active, perpendicular dichroism. Weak Raman lines and strong IR bands.
$9E_2$: Raman active, weak Raman lines, depolarized. For PTFE, calculations show that the E_2 modes lie close to the frequency of A modes.

Thus, 13 modes are unique to the Raman, while only 3 modes are unique to the infrared, with 8 modes common to both.

The Raman spectrum of PTFE appears in Figure 9.6 (18). The A_1 modes are marked, the A_2 are three unique IR bands (516, 636, and 1210 cm^{-1}), and the E_1 modes in the Raman spectrum are identified by their coincidence with IR bands. No major frequency shifts occur upon cooling the sample below 19°, in agreement with calculations that show small differences in expected frequencies for 13 CF_2's per six turns of the helix and the 15/7 helix. However, since considerable frequency differences would be expected in the vibrational pattern for the planar zigzag structure, this conformation may be excluded from consideration.

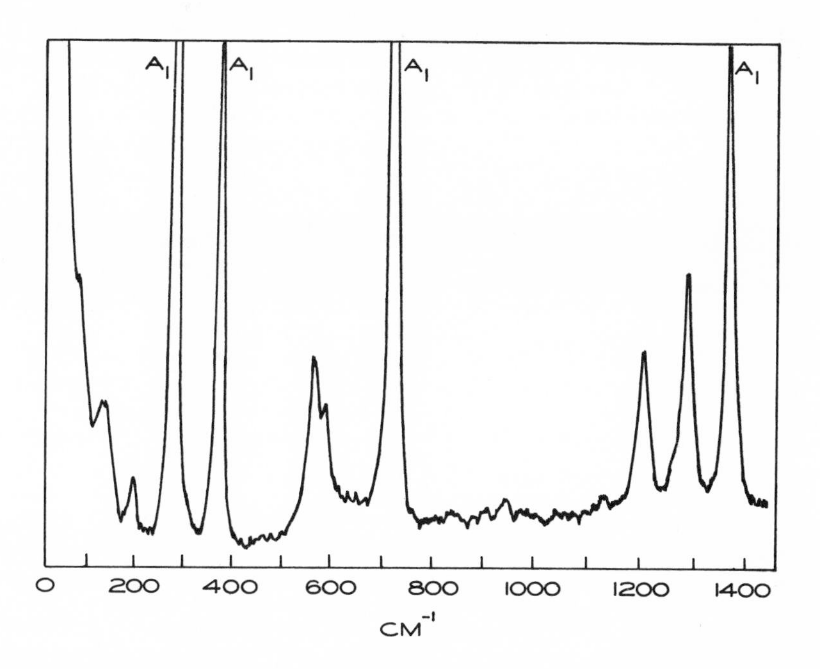

Figure 9.6. Raman spectrum of polytetrafluoroethylene (18).

9.4.2 Nylons

Differing from polyhydrocarbons and other relatively nonpolar polymers, the association of the individual chains of nylon in the solid phase mainly involves the hydrogen-bonded interaction of a C=O group on one chain with an N-H on another. A regular lattice of hydrogen-bonded amide groups results, and the methylene sequences simply bridge these in whatever configurations they are constrained. Although such chain-chain interactions greatly influence the mechanical properties of the polymers, the methylene configuration sequence also exerts an effect on these properties. Information pertaining to the skeletal orientations can be gained from Raman data, but none of the other structural techniques available to polymer chemists is likely to be of assistance in this particular area.

The IR spectrum of a sample thought to be a nylon can confirm that the material is a polyamide, but distinguishing one nylon from another is a difficult task because the IR spectrum is dominated by bands arising from vibrations of the amide groups. Since the strong bands in the Raman spectra of various nylons are due to the backbone methylene sequences, differentiation is readily achieved (Figure 9.7) (10).

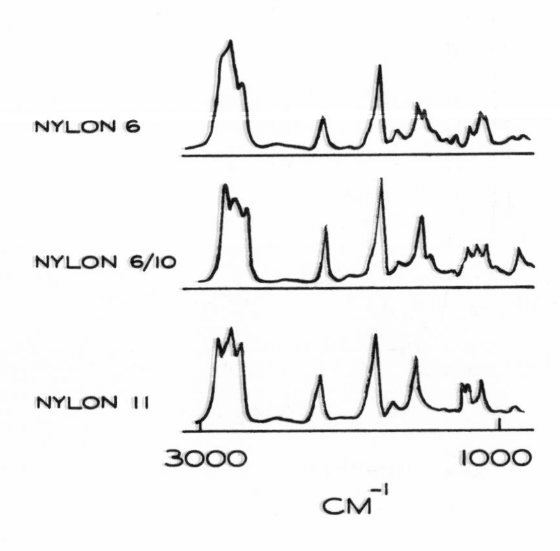

Figure 9.7. Raman spectra of some chemically similar nylons (10).

In contrast to IR, no sample preparation is necessary in the Raman method. Excepting Nylon 6 and Nylon 6/6, one can distinguish among Nylon 6, 11, 12, 6/6, 6/10, and 6/12. However, the IR spectra of Nylon 6 and 6/6 are sufficiently different to permit their identification (9).

9.5 CHAIN CONFORMATION IN SOLUTION

9.5.1 Polyethylene Glycol (PEG)

Polymer structures often change upon dissolution or with variations in the pH, salt content, or ionic strength of the solution. Raman spectroscopy is outstanding in aqueous solution because water is a very weak scatterer except for 1650 and 3600 cm^{-1} bands. The spectra of PEG in aqueous and chloroform solutions are shown in Figure 9.8 (4).

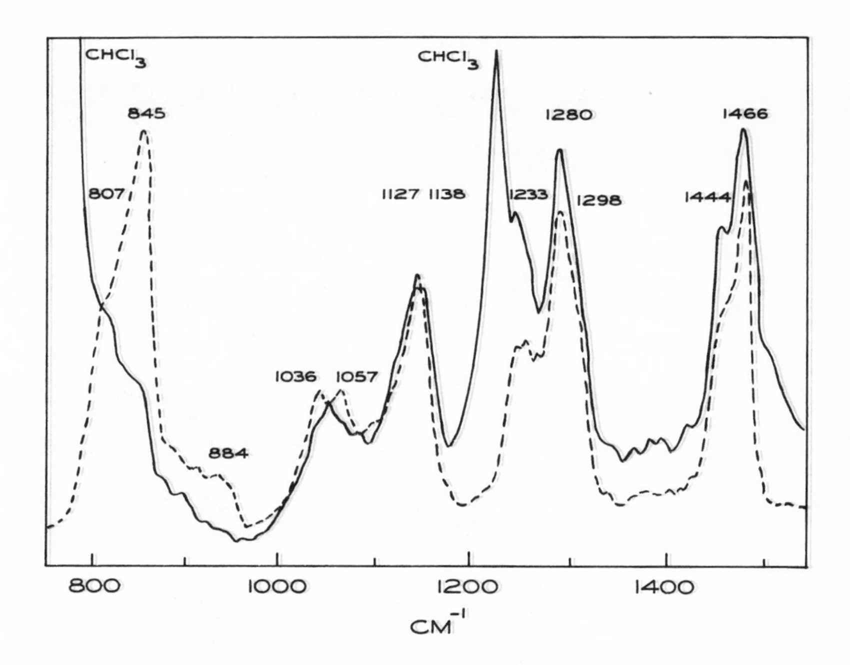

Figure 9.8. Raman spectra of PEG in aqueous solution and chloroform (4).

The profile of PEG in aqueous solution resembles that of the solid polymer, while it is quite different from the chloroform solution which, in turn, is similar to the spectrum of PEG melt. The data are in agreement with IR and NMR results which suggest that the conformation of PEG in aqueous solution retains to a large extent the TGT conformation of the COCCOC sequence. Of the three Raman lines occurring in the methylene rocking region (884, 845, and 807 cm^{-1}), the line ascribed to the TGT structure (845 cm^{-1}) is appreciably stronger. The observation that the spectra of the molten polymer and the chloroform solution are very similar indicates a disordered structure as a result of additional rotational isomers, since additional Raman lines occur compared to the crystalline spectrum (4).

REFERENCES

1. R. Signer and J. Weiler, Helv. Chim. Acta 15, 649 (1932).
2. M. Patel, Current Sci. (India) 18, 136 (1949).
3. R. F. Schaufele, J. Opt. Soc. Am. 57, 105 (1967).
4. J. L. Koenig, Applied Spec. Rev. 4 (2), 233 (1971).
5. S. W. Cornell and J. L. Koenig, Macromolecules 2, 540 (1969).
6. S. W. Cornell and J. L. Koenig, Macromolecules 2, 546 (1969).
7. J. L. Koenig, M. M. Coleman, J. R. Shelton, and P. H. Starmer, Rubber Chem. and Tech. 44, 71 (1971).
8. J. L. Koenig and M. Meeks, J. Polymer Sci. 9, 717 (1971).
9. G. Schreir and G. Peitscher (private communication, 1972).
10. M. J. Gall and P. J. Hendra, The Spex Speaker XVI, No. 1 (1971).
11. A. J. Melveger, J.Polymer Sci. A-2 10, 317 (1972).
12. G. Schreir and G. Peitscher, Z. Anal. Chem. 258, 199 (1972).
13. S. Krimm, A. R. Berens, U. L. Folt, and J. J. Shipman, Chem. Ind. (London), 1512 (1958).
14. S. Krimm, A. R. Berens, U. L. Folt, and J. J. Shipman, Chem. Ind. (London), 433 (1959).
15. G. Natta and P. Corrandini, J. Polymer Sci. 20, 251 (1956).
16. J. L. Koenig and D. Druesdow, J. Polymer Sci. A7, 1075 (1969).

17. J. R. Nielsen and A. H. Woolett, J. Chem. Phys. 26, 1391 (1957).
18. J. L. Koenig and F. J. Boerio, J. Chem. Phys. 50, 2823 (1969).

Chapter 10

BIOLOGICAL MATERIALS

10.1 INTRODUCTION

Raman and IR spectroscopy are excellent complementary probes for the study of macromolecules exemplified by biopolymers and proteins. It must be recognized that, although the Raman effect is similar to IR because it deals with molecular vibrations, the Raman spectrum of a compound can be more useful to biological studies than merely supplying more complete vibrational information. For example, spectra obtained from solids can be compared with aqueous solution spectra in order to determine the effect of solvent molecules on conformation. In addition, isotopic exchange data can be gained by comparing spectra in H_2O and D_2O. Sample presentation techniques favor Raman over IR, for example, materials can be studied intact and in the natural state without the influence of foreign materials such as salts or oils, which may affect the conformation of biological macromolecules.

Vibrational spectroscopy is an established physical method for studying biopolymers, but until recently the vibrational information has been obtained chiefly by IR spectroscopy. Several pre-laser Raman studies were conducted in aqueous solutions between 1936 and 1958. The increase in the application of Raman spectroscopy to biological areas was due to the availability of laser sources, particularly the Ar^+ laser. Fluorescence is the major factor for the delay in a more general use of this spectral technique, but application of the various methods to reduce the extent of fluorescence has ameliorated the problem, and increased activity can be expected in this field.

10.2 AMINO ACIDS

Raman studies concerned with distinguishing individual amino acids have been applied to the analysis of proteins. More than thirty years ago it was recognized that ionization of the amino acid NH_2 group influenced the frequency of the carboxylic acid C=O vibration (Table 10.1) (1).

TABLE 10.1. Raman C=O Frequencies for Carboxylic Acids (1)

Compound type	$\nu_{C=O}$ (cm^{-1})
R-COOH (solid)	1670
R-COOH (aqueous solution)	1720
R-C·(NH_3^+) - COOH	1740
R-C·(NH_2) - COO^-	1570-1600 ∿1400

The COOH frequency is observed at 1740 cm^{-1}, 20 cm^{-1} higher than that of aliphatic acid carbonyl vibrations. The amino acid carbonyl band vanishes upon ionization of the acid group, being replaced by two bands: an asymmetric stretching mode (1570-1600 cm^{-1}) and a symmetric mode (near 1400 cm^{-1}). The latter band is very intense in the anionic and dipolar forms of all amino acids and essentially is unaffected in position or intensity by deuteration or conjugation of the amino group.

10.3 BIOPOLYMER CONFORMATIONS

In order to adequately understand the stabilization forces that maintain the numerous conformations found in polypeptides and proteins, a definition of "conformation" is necessary. The four terms employed for defining protein structure are: (a) primary structure--the sequential arrangement of the amino acids; (b) secondary structure--the organization of these backbones into three-dimensional structures, such as helices and sheets; (c) tertiary structure--the total folding of this secondary structure; and (d) quaternary structure--the arrangement of the different chains with respect to each other. A conformational change is defined as any alteration in the ordered structure of a polypeptide chain or protein, that is, involving a change in secondary, tertiary, or quaternary structures. The determination of folding of the polypeptide chains is one of the keys to an understanding of protein structure.

The conformation of the amide group is of overriding importance in determining the backbone structure of polypeptides and proteins. The partial double bond character of the CO-NH bond in polypeptides and proteins results in a planar conformation of the amide groups. In addition to planarity, the amide bonds in linear or open-chain polypeptides (2) and proteins (3) adopt the *trans* conformation exclusively, while the *cis* planar form has been proposed for some residues of fibrous proteins (4). Assignments of the various amide bands appear in Table 10.2. The major difference between the Raman and IR spectra reside in band intensities, that is, some bands that are weak or absent in the Raman effect may be readily observed in the IR, and vice versa. Provided that the amide vibrational modes arise from localized vibrations of the peptide group, the characteristic bands may be correlated with conformation type.

10.4 AMIDE I AND AMIDE II BANDS OF POLYPEPTIDES

The relationship for Raman and IR amide I and II bands of the crystalline regions of polypeptides arises from in-phase motions of corresponding groups in various cells, taking into account vibrational interactions transmitted through hydrogen bonds and among peptide groups in the chain (5) (Figure 10.1),

$$\nu(\delta, \delta') = \nu_o + \Sigma D_1 \cos(s\delta) + \Sigma D_1' \cos(s'\delta'), \quad (10.1)$$

where ν_o is the unpertubed frequency of the amide I

TABLE 10.2. Vibrational Modes of the *Trans* Secondary Amide Group

Band Terminology	Approximate $\nu_{C=O}$ (cm^{-1})	Assignment
Amide I	1650[a]	Primarily C=O stretch.
Amide II	1550[b]	Primarily N-H in-plane bending with a C-N stretching contribution.
Amide III	1250[c]	Primarily C-N stretching with N-H in-plane bending. (The coupling phases are opposite to Amide II).
Amide IV	630[d]	O=C-N bending.
Amide V	700[d]	N-H out-of-plane bending.
Amide VI	600[d]	C=O out-of-plane banding.
Amide VII	200[d]	Internal rotation about the peptide C-N bond related to the barrier hindering internal rotation.

[a]*cis* - 1650
[b]*cis* - 1450
[c]*cis* - 1350
[d]Weak in Raman

O=C-N(-H): *trans* planar

O=C-N(-H): *cis* planar

(or II) modes, and δ and δ' are the phase differences between intrachain and interchain vibrational motions of adjacent groups, respectively. The unperturbed frequency is corrected by the constants D_1 and D'_1, which are determined by the potential and kinetic energy interactions associated with hydrogen bonds between s^{th} neighbors in and between the chains, respectively.

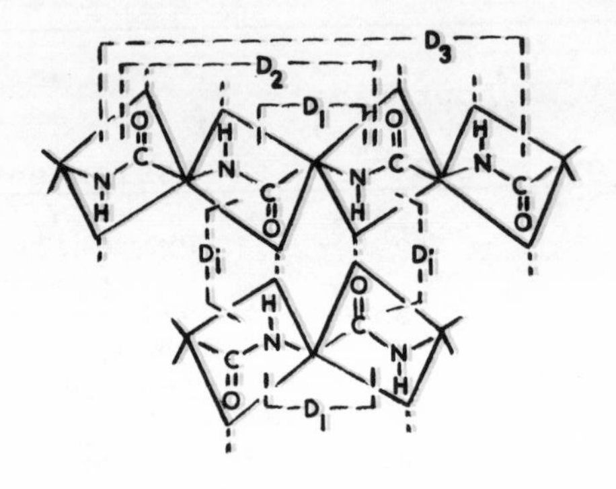

Figure 10.1. Vibrational interactions among peptide groups (5). [Courtesy of Marcel Dekker, Inc.]

10.4.1 The β Conformation

A common chain conformation of polypeptides is the β conformation, in which the polypeptide chain is nearly fully extended. An antiparallel pleated sheet structure accommodates the hydrogen bonding. The four vibrational modes of this form are shown in Figure 10.2 (5). The unit cell contains four peptide groups: two groups from one chain and the other two from an adjacent chain. Four bands are expected to appear in the vibrational spectra corresponding to the in- and out-of-phase motion of the amide groups on adjacent chains and on the same chains. (The arrows represent the components of the transition moments of the peptide group in the plane of the paper. The plus and minus signs represent the components of the transition moments perpendicular to the plane of the paper; the former pointing upward, the latter pointing downward.)

The frequencies of the vibrational modes are derived from Equation 10.1 and are shown in Equations (10.2a-(10.2d):

$$\nu_{\parallel}(0, 0) = \nu_0 + D_1 + D_1' \quad (10.2a)$$

$$\nu_{\parallel}(0, \pi) = \nu_0 + D_1 - D_1' \quad (10.2b)$$

$$\nu_{\perp}(\pi, 0) = \nu_0 - D_1 + D_1' \quad (10.2c)$$

$$\nu_{\perp}(\pi, \pi) = \nu_0 - D_1 - D_1' \quad (10.2d)$$

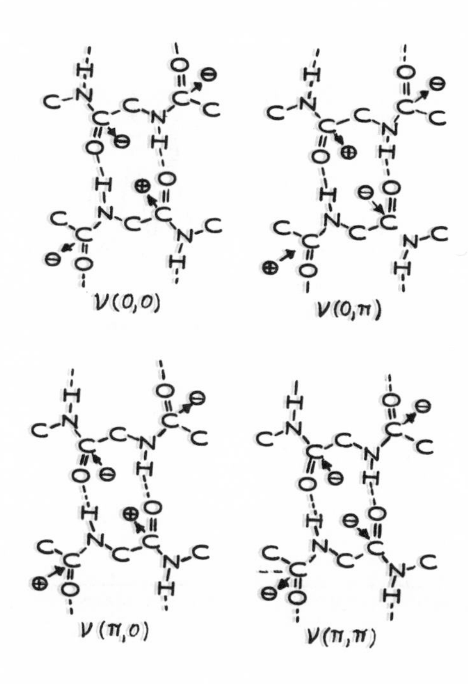

Figure 10.2. The vibrational modes of the antiparallel chain β-pleated sheet (5). [Courtesy of Marcel Dekker, Inc.]

Corresponding to Equation 10.1, the first term in parentheses in the above equations is the phase angle δ (intrachain), and the second term is the phase angle δ' (interchain). The frequency notations $\nu_{\parallel}$ and $\nu_{\perp}$ signify that the transition moments of the vibrations are parallel and perpendicular, respectively, to the fiber axis. These equations are useful in making vibrational assignments for the β conformation. All of the bands are Raman active; all but one, $\nu_{\parallel}(0,0)$, are IR active. The Raman spectra of several polypeptides display a strong band in the amide I region at approximately 1670 cm^{-1} that probably originates from the $\nu_{\parallel}(0,0)$ vibrational mode (6,7,8). The Raman spectrum of poly-L-serine is shown in Figure 10.3 and pertinent assignments are listed in Table 10.3 (6).

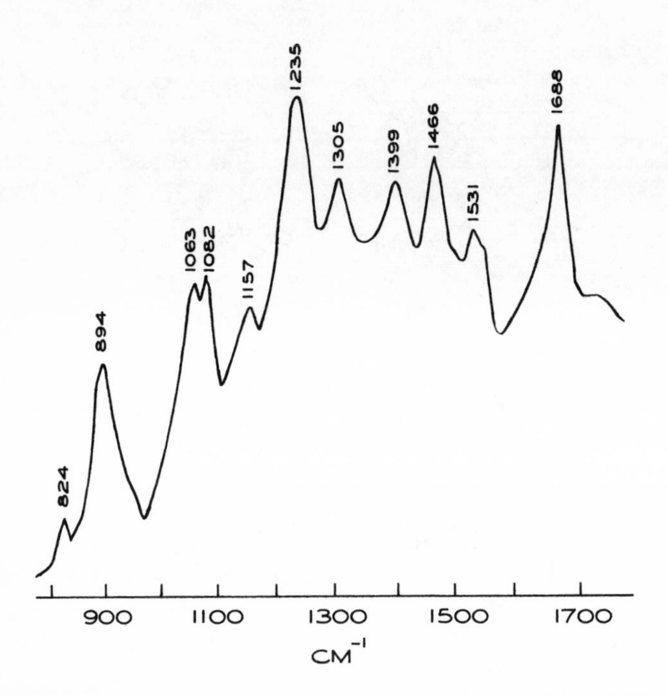

Figure 10.3. Raman spectra of poly-L-serine (6).

10.4.2 α-Helical Conformation

Many polypeptides and proteins exist in the α-helical conformation, in which there are 18 amino acid residues in five turns (3.6 per turn) and interchain hydrogen bonds are found between each pair of every third neighbors (Figure 10.4, Equation 10.3) (5).

$$\nu(\delta) = \nu_o + D_1 \cos\delta + D_3 \cos 3\delta \qquad (10.3)$$

Spectral bands arise either from vibrations with a phase difference $\delta = 0$ (parallel band) or from degenerate vibrations with a phase difference $\delta = \theta$ (perpendicular band), where θ is the phase difference between corresponding atoms in adjacent repeat units. The Raman and IR active vibrations of the α helix correspond to $\theta = 0$ and $\theta = 2/n$, where $n = 3.6$. Since some of these bands that are too weak to be detected in the

TABLE 10.3. Raman and Infrared Frequencies for Poly-L-Serine (6)

Raman	IR	Group	Tentative Assignment
1668 vs	1695 wsh		
1668 vs		skeletal	Amide I
	1621 s		
1531 vw	1537 wsh	skeletal	Amide II
	1512		
1466 s	1460 wsh	residue	
1399 m	1397 w	residue	
	1321 vvw	residue	
1304 m		skeletal	
	1288 vvw		
1235 vs	1235 m	skeletal	Amide III
	1200 wsh		
1157 m			
	1141 m		
1082			
1063 s	1056 s	residue	Primary alcohol group
	936 m		
894 s	894 w		
	861 w		
824 w			
	525 vw		

Raman spectra may be observed in the IR spectra, and vice versa, both measurements are required for a complete analysis. This statement also applies to other conformational types.

A strong Raman band near 1655 cm^{-1} is observed in the spectra of polypeptides adopting the α-helical forms (6,9,10), approximately 15 cm^{-1} lower than the frequency for the β conformation. The Raman spectrum of poly-L-leucine and band assignments (6) appear in Figure 10.5 and Table 10.4. Raman spectra in the solid state, in water, and in aqueous salt solutions indicate that, under all these conditions, block copolymers of L-alanine and D,L-lysine are sensitive to the conformational states of the polymers (11). Spectral data suggest that both blocks consist of mixtures of α helix and random coil in the solid state.

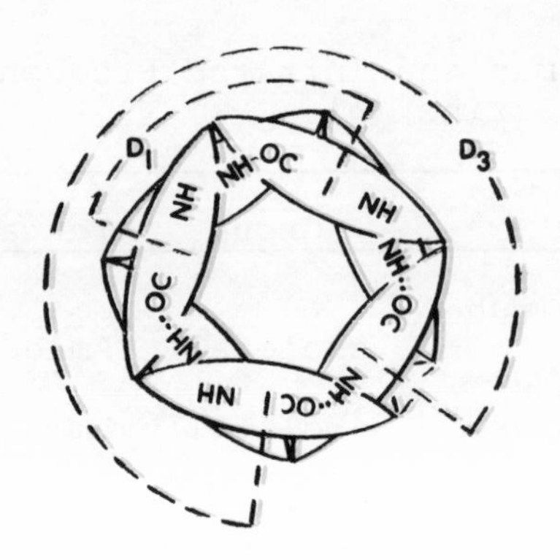

Figure 10.4. Vibrational interactions for the α helix (5). [Courtesy of Marcel Dekker, Inc.]

The poly-D,L-lysine exists exclusively in the random coil form in aqueous solution, but the poly-L-alanine remains as a mixture of helical and coil conformations. In addition, the spectra in aqueous salt solutions indicate the presence of interactions not present in salt-free water. This may be explained by the shielding of the charges of the poly-D,L-lysine end blocks in salt-free water, permitting the central poly-L-alanine block to adopt a hairpinlike conformation (with the helices folded back and forth) that is stabilized by side chain methyl-methyl hydrophobic bonding.

10.4.3 Raman and IR Amide I Vibrations

It is of interest to note that Raman and IR techniques do not detect the same amide I vibrations in both peptide homopolymers and proteins (12,13). A comparison of band locations for various materials is listed in Table 10.5. The discrepancies may be due to excitation of coupled vibrations between adjacent peptide units. In the case of denatured insulin, one coupled mode gives rise to an intense Raman band at 1673 cm^{-1} while the other coupled mode is observed as an intense IR band at 1637 cm^{-1}.

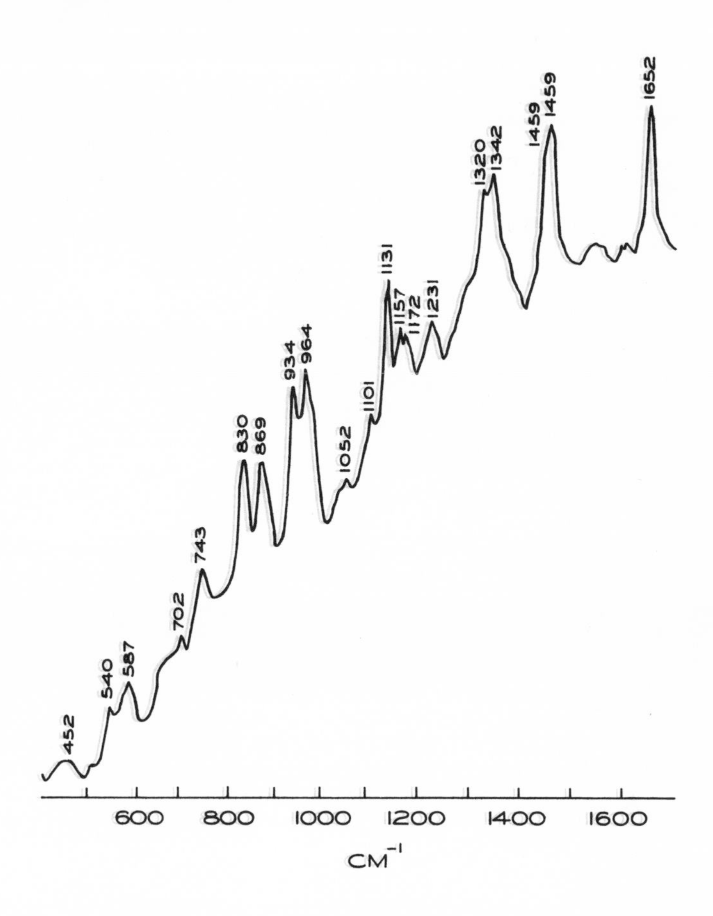

Figure 10.5. Raman spectrum of poly-L-leucine (6).

10.5 THE AMIDE III BAND

Studies of the Raman spectra of synthetic polypeptides indicate that the amide III as well as the amide I frequencies of α-helical, random-coiled, and β-pleated sheet forms are quite different. For example, the α-helical poly-L-alanine (14) has its amide III vibrational mode at 1264 cm^{-1}, the random-coiled poly-L-glutamic acid (15) at 1248 cm^{-1}, and the antiparallel β polyglycine I (7) at 1234 cm^{-1}.

TABLE 10.4. Raman and IR Frequencies for Poly-L-Leucine (6)

Raman	IR	Group	Tentative Assignment
1652 vs	1650 vs	Skeletal	Amide I
1547 vw	1540 vs		Amide II
	1506 vw	Skeletal	
	1500 vw		
	1466 m		
1459 vs		Residue	
	1449 w		
	1436 w		
	1382 m	Residue	
	1364 s		
1342 vs	1344 vs	Residue	
1320 vs		Residue	
	1291 w		
	1253 m		Amide III
1231 m		Residue	
	1218 m		
1172 m	1168 s		
1157 m		Residue	
1131 s		Residue	
	1122 s		
1101 vw		Residue	
	1094 m		
1052 w		Residue	
	1042 w		
	1020 w		
	973 w		
964 vs		Residue	
934 m	936 vs		
	920 vw	Residue	
869 s	869 s	Residue	
830 s	829 w	Residue	
	722 w		
743 m		Skeletal	
702 w		Skeletal	
	694 m		
	657 vw	Skeletal	Amide V
	614 m		
587 m	582 w	Skeletal	
540 vw		Skeletal	
	471 s	Skeletal	
452 w		Residue	

TABLE 10.5. Raman and IR Amide I Vibrations in Some Peptide Homopolymers and Proteins

Peptide or Protein	Amide I Band (cm^{-1})	
	Raman	IR
Polyglycine I (7)	1674 (strong)	1636 (strong) 1685 (medium)
Glucagon fibrils (12)	1672 (strong)	1630 (strong) 1685 (weak)
Insulin (denatured fibrous) (13)	1673	1637

10.6 PROTEINS

Although X-ray and electron diffraction studies can yield detailed structural information, biological activity does not exist in the solid state. A technique is required to ascertain whether or not the structural features of a protein are the same in aqueous ionic solution as in the solid phase. Raman spectroscopy is such a technique. Furthermore, the involvement of protein molecules in biological processes, such as reactions with enzymes, often can be determined by Raman spectroscopy. The usefulness of Raman spectroscopy has been amply demonstrated in protein studies and there is little doubt that future investigations will be aided greatly by the technique.

Generating Raman spectra of proteins posed a formidable challenge until the Argon ion laser became available. Presenting the sample to the instrument immediately subsequent to preparing it in a cold room significantly reduces the fluorescence problem (16). Protein studies using spectroscopic methods are hampered by low concentrations of these materials in aqueous solution. Water is normally an excellent medium because it is a relatively weak scatterer, but since proteins exist in biological fluids at levels of 0.1-1%, Raman water bands become quite strong under the instrumental conditions necessary to observe solute lines. However, computer techniques often allow the use of low sample concentrations in water (17).

The Raman spectral quality of proteins is inferior to that of polypeptides. This may be attributed in part to the fact that Raman scattering is approximately proportional to the mass of the molecule per unit volume, while Rayleigh scattering increases with increasing size of the molecule. Furthermore, since many of the weak Raman bands overlap, the spectra generally become diffuse. The spectra of crystalline lysozyme, pepsin, and α-chymotrypsin were recorded several years ago (18). These enzymes are weak scatterers and have extremely broad bands, but enough minor differences were observed to indicate at that time that Raman spectroscopy was a useful method of identification.

10.6.1 Lysozyme

Lysozyme, a mucolytic enzyme possessing antibiotic properties, is one of the most thoroughly investigated enzymes. Its primary structure and three-dimensional structure in the native state are as well known as its active site and details of its enzymatic action.

The Raman spectrum of lysozyme has been interpreted by superimposing Raman spectra of the twenty different kinds of amino acids comprising the molecule (15). The correspondence is quite acceptable (Figure 10.6). The Raman spectrum of solid native lysozyme is quite similar to spectra of aqueous solutions in the 1-10% range (19).

The optical inhomogeneity produced by thermally denaturing lysozyme militates against a Raman study of structural changes in this enzyme, but chemical denaturation with 6M LiBr can be accomplished without obscuring the spectrum (19). The Raman bands of native lysozyme in water are mainly those of the amino acid side chains, but two bands characteristic of the peptide backbone are centered at 1660 cm^{-1} (amide I) and 1262 cm^{-1} (amide III). The latter contains at least three components that appear as broad maxima at 1274, 1262, and 1240 cm^{-1} [Figure 10.7(A)], the entire group shifting to 940 cm^{-1} upon deuteration. These peaks are ascribed to amide III vibrations in segments of the

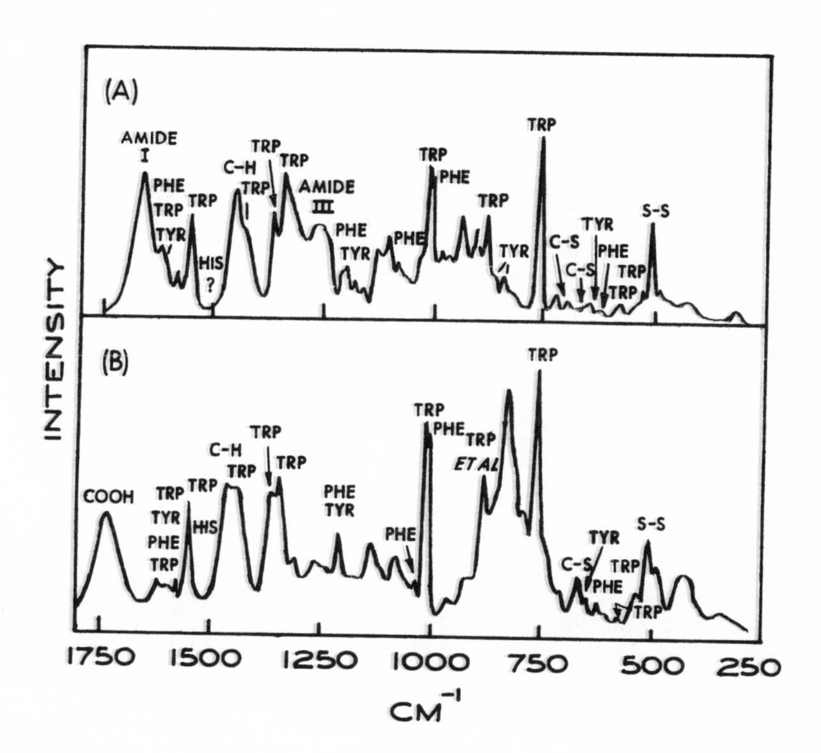

Figure 10.6. Raman spectrum of lysozyme (a) in water, (b) spectra resulting from superposition of spectra of constituent amino acids (15).

backbone in the α helix, β-pleated sheet, and random coil conformations. The shift in the center of gravity of the amide III band profile from above 1260 cm^{-1} to near 1245 cm^{-1} [Figure 10.7(B)] is interpreted to mean that the ordered structure in the protein backbone is removed, leaving a random coil structure only.

The denaturation of lysozyme has been studied by temperature and by chemical reagents that cleave the S-S bonds (20,21,22). Under those conditions in which the enzyme denatures reversibly (pH 5 and 75°), the amide III region is almost identical with Figure 10.7(A). From this it may be concluded that the peptide backbone of lysozyme retains essentially the same amount and kind of ordered structure as at room temperature.

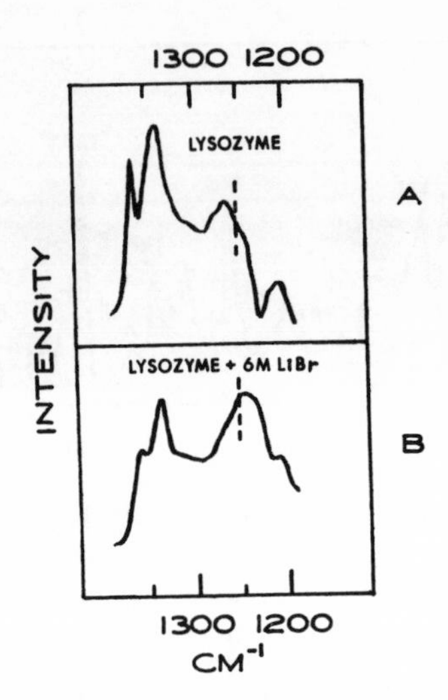

Figure 10.7. Raman spectra of 5% aqueous lysozyme (pH 4.2, t = 25°). The vertical dashed line indicates 1250 cm^{-1} (19).

On the other hand, when the disulfide bonds are ruptured chemically and the resultant sulfydryl groups blocked with acrylonitrile, the amide III region very closely resembles Figure 10.7(B). Expectedly, the ordered structure of lysozyme is destroyed when the disulfide bonds are broken (19). The S-S stretching band at 504 cm^{-1}, arising from the four cystine residues per molecule, is strikingly temperature dependent (22). A decrease in intensity occurs when lysozyme is heated in aqueous solution, essentially reaching a zero value above 76°. Neither the frequency nor the band width changes during heating of the enzyme. It may be assumed, then, that all the disulfide bonds are broken in the denatured state above 76°. In contrast, NMR investigations (23,24) indicate that the disulfide linkages remain unaffected by this denaturing process. However, the NMR technique does not directly probe the S-S bondings.

10.6.2 Insulin

Insulin is a hormone concerned in the regulation of

carbohydrate metabolism. It can be transformed from globular to fibrous form by heating in dilute HCl solution, with a concomitant loss in activity and solubility (25). Both the activity and solubility can be recovered by treating the inactivated protein with 10% aqueous phenol solution (26). The nature of this transformation has been studied by various techniques (25).

Comparison of the Raman spectra of native zinc-insulin crystal with denatured insulin reveals dramatic changes [Table 10.6, Figures 10.8(A) and 10.8(B)], indicating that the structure and conformation have undergone considerable changes during denaturation (13,27). The amide I band at 1664 cm^{-1} is sharpened and shifted to 1673 cm^{-1} upon denaturation, suggesting that the denatured molecule exists in a β conformation. This spectral change may be ascribed to alterations in inter- and intrachain coupling between adjacent peptide units and to a weakening, accompanied by a greater uniformity, in hydrogen bonding.

Three resolved peaks are observed in the amide III region (1239, 1270, and 1288sh cm^{-1}), whose center of gravity is markedly shifted from 1260 to 1230 cm^{-1} on denaturation. The 1270 cm^{-1} band in the native insulin spectrum is assigned to the α-helical form and the one at 1239 cm^{-1} to the random coil structure with only a small contribution from the β structure (13,27). The spectral feature at 1230 cm^{-1} for the denatured sample agrees with the amide III frequency for model compounds with known β structure (see p.262).

In native insulin the S-S frequencies of the three disulfide linkages are observed as an unresolved broad band centered at 515 cm^{-1}, while the C-S frequencies appear at 668 and 678 cm^{-1}. Interesting and informative spectral changes occur when insulin is denatured. The peak intensity of the S-S frequency increases and the scattering at 668 cm^{-1} shifts to 657 cm^{-1} with a considerable decrease in intensity. The band at 678 cm^{-1} undergoes but a slight shift to 680 cm^{-1} with no significant loss in intensity. Since the intrachain disulfide bond may have a local geometry different from that of the interchain disulfide bond, this observation suggests that the geometry of the two interchain disulfide bonds in the denatured insulin is different from that in the native state and that the intrachain one remains in nearly the same conformation upon denaturation (27).

TABLE 10.6. Raman Band Assignments for Beef Insulin (27)

ν (cm^{-1}) Native (crystals)	Denatured (solid)	Tentative Assignments
	265	Skeletal bending
333	325	
410	420	
467	460	
495	480	
515	513	ν(S-S)
	532	Skeletal bending
563		
624	624	Phe
644	644	Tyr
668	657	ν(C-S) of the C-S-S-C group
678	680	
725		Skeletal bending
747	737	
770		
814		
832	830	Tyr
854	853	
900	882	ν(C-C)
934	922	
946		
963	956	
1004	1004	Phe
	1020	ν(C-N)
1032	1032	Phe
	1057	ν(C-N)
1112		
1128	1127	
1162	1161	

TABLE 10.6. (Continued)

ν (cm^{-1}) Native (crystals)	Denatured (solid)	Tentative Assignments
1177	1175	Tyr
1212	1214	Tyr & Phe
	1227 1252	Amide III (β-structure)
1239		Amide III (random-coil)
1270		Amide III (α-helical)
1288		Amide III (α-helical)
1322 1344	1327 1343	CH deformation
1367		
	1407	
1425	1422	Symmetrical CO_2^- stretching
1450 1462	1450 1462	CH_2 deformation
1587	1587	Phe
1607	1607	Phe & Tyr
1615	1615	Tyr
1662		Amide I (α-Helical structure)
	1673	Amide I (β-structure)
1685		Amide I (random-coil)
	1735	-COOH

The intensity decrease of the bands at 832 and 854 cm^{-1} (assigned to the ring vibrations of the tyrosine residues) may be due to the changes of the environments of these rings as a result of extensive unfolding of the backbone. In addition, the Raman spectral similarity among amorphous, aqueous, and native zinc-insulin crystals indicates that they exist in the same conformation (27).

10.6.3 Glucagon

Glucagon is a polypeptide hormone of pancreatic origin

that is capable of undergoing the α-helix--random-coil--β-pleated sheet transition. Prior to a recent Raman study (12), it has been reported that glucagon existed in 75% α-helical structure in crystals (28), and that in freshly prepared acidic solutions it was predominantly in the form of a random coil (29). This acidic solution is gradually converted into a gel at 28°, consisting of antiparallel chains (30).

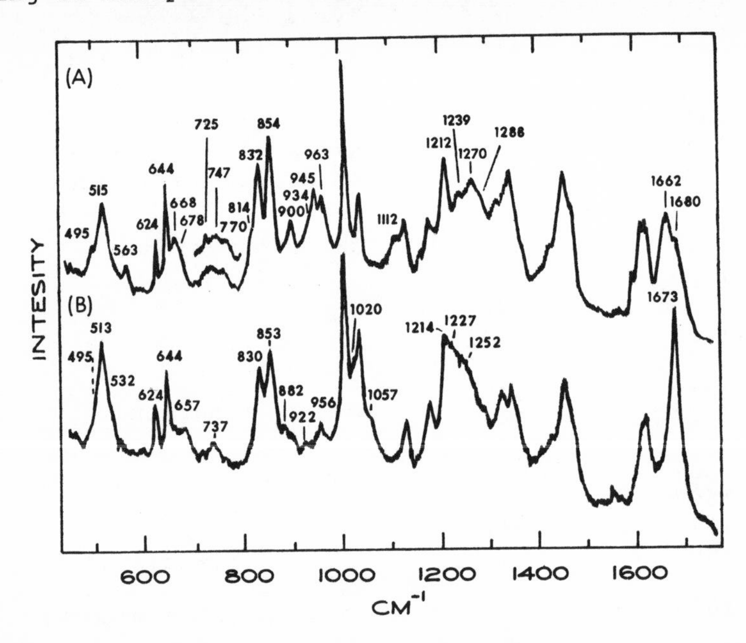

Figure 10.8. Raman spectra of beef insulin. (a) spectrum of native zinc-insulin crystalline powder; (b) spectrum of denatured insulin (heat-precipitated) (27).

The Raman spectra of glucagon in crystals, freshly prepared aqueous solution (pH 2.25), and gels are presented in Figures 10.9(A) and 10.9(C) (12). A stepwise decrease in frequency of the amide III lines is observed from 1266 (α helix) to 1248 (random coil) to 1232 cm^{-1} (antiparallel β). The amide I band of crystalline glucagon [Figure 10.9(A)] appears at 1658 cm^{-1}. The shoulder near 1685 cm^{-1} may arise from the unsolvated random-coiled segments of the protein (about 25%). The intense water band near 1640 cm^{-1} [Figure 10.9(B)] obscures the amide I frequency. Upon gel formation [Figure 10.9(C)], however, the amide I frequency is seen as a very strong band at 1672 cm^{-1} on the sloping water background. The two bands at 1658 and 1672 cm^{-1} are in agreement with the corresponding frequencies of α-helical poly-L-alanine and antiparallel β-polyglycine I. The amide I and amide III Raman frequencies obtained on glucagon fibrils in the solid state are the same as those found in Figure 10.9(C), indicating that the antiparallel β structure of glucagon gels remains in the solid state (12).

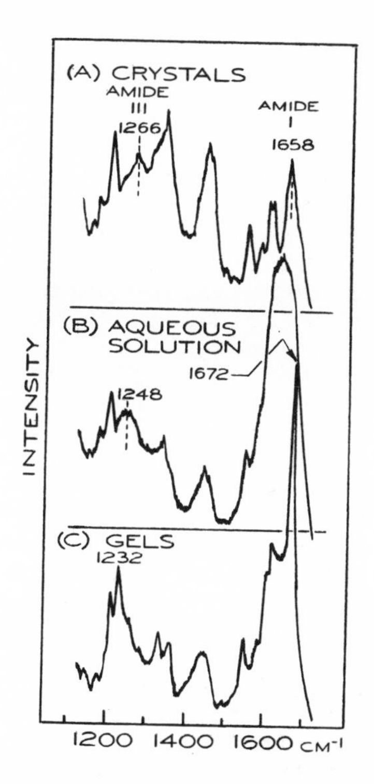

Figure 10.9. Raman spectra of Glucagon in various states of aggregation (12).

10.6.4 Collagen

The Raman spectrum generated on a chip of defatted ox tibia shows intense bands originating from the symmetric phosphate modes (31). In addition, several of the bands present are due to backbone and side chain vibrations of collagen, a protein contained in connective tissue and bone. There is evidence that both proline and phenylalanine in the collagen contribute to the spectrum. A Raman spectrum obtained from tooth dentin is similar to this bone spectrum in general form (32). These results demonstrate the feasibility of obtaining Raman spectra for fibrous proteins and indicate the possibility of comparing normal and pathological tissues as well as conducting aging studies.

10.6.5 Native Bovine Serum Albumin and β-Lactoglobulin

Bovine serum albumin is a large protein of molecular weight 64,000 cross-linked by 16 disulfide bonds. Bovine β-lactoglobulin has a monomeric molecular weight of 18,000 and is cross-linked with two S-S bonds. The Raman spectra of these proteins possessing widely differing secondary structures have been interpreted with the aid of amino acid data (Table 10.7) (33). A few interesting facts emerge from a study of the spectra. Bands due to tyrosine appear for bovine serum albumin at 820 and 852 cm^{-1}; in β-lactoglobulin at 832 and 860 cm^{-1}. These significant differences reflect different environments of the tyrosine residues in the molecule that result from differences in the hydrogen bonding of the hydroxyl group in tyrosine. The intensity ratio of the S-S/C-S stretching bands (see p.233), is approximately the same as that found in lysozyme which suggests that the C-S-S-C dihedral angles are on the average similar to those in lysozyme.

The C-S stretching frequencies of the methionine residue depend on its conformation (15). Two bands are expected for the *trans* form at 655 and 724 cm^{-1}, while the *gauche* conformer has two bands at 655 and 700 cm^{-1}. In β-lactoglobulin, with four methionine residues, a weak band is observed at 703 cm^{-1} along with one at 650 cm^{-1} but no band appears near 724 cm^{-1}. This suggests that most of the methionine residues are in the *gauche* form for this protein.

The amide III band of β-lactoglobulin is shown at 1242 cm^{-1} with a shoulder at 1262 cm^{-1}. In bovine serum albumin a shoulder occurs at 1280 cm^{-1} and a broad band at 1250 cm^{-1}. In each case the observation of these two frequencies suggests two structural components of substantial abundance in the protein backbone. The 1262 cm^{-1} band in β-lactoglobulin has been attributed to β-structure and the 1242 cm^{-1} band to the random coil conformation (33). The former assignment is not in agreement with β forms reported for some other proteins, for example, insulin (see p.272) and glucagon (see p.275). In bovine serum albumin the shoulder at 1280 cm^{-1} is indicative of α helix, while the broad 1250 cm^{-1} band is assigned to the random-coil form.

10.6.6 Ribonuclease A (RNase A)

Most proteins are not denatured by lyophilization (freeze drying), and they may be stored in the dry state for long periods of time without deterioration. Part of the water probably is tightly bonded to protein molecules and cannot be released even on drastic drying (34). The lyophilization process appears to remove unbonded and loosely bonded water molecules from the frozen sample, bringing about the replacement of a number of protein-solvent contacts by protein-protein contacts. The Raman spectrum of RNase A in solution and freeze dried powder form show differences which reflect conformational changes (35). Upon dissolution the intensities of the 644 and 852 cm^{-1} bands due to the ring vibrations of tyrosine residues decrease significantly relative to the 832 cm^{-1} band. The observed intensity variations are attributed to changes in the local environment of the three tyrosines having an abnormally high pK_A value; this environmental change probably resulting from conformational changes in the vicinity of these groups. The remaining three tyrosines of RNase A, exhibiting normal pK_A values, are not involved.

10.6.7 Cobramine B

Cobramine B, a small basic protein from cobra venom, has been selected as a model for studying the scattering intensity of tyrosyl ring vibrations in the Raman spectra of proteins (36).

TABLE 10.7. Raman Spectra of Native Bovine Serum Albumin and β-Lactoglobulin in Water

Bovine Serum Albumin	β-Lactoglobulin	Tentative Assignment
468*	469	H_2O
	490 sh	
	510 (2)	ν(S-S)
	578 (0)	Trp
	605 (1)	
623 (2)	628 (2)	Phe
650 (1)	650 (1)	ν(C-S) + Tyr
675 (1)		
695 (0)	690 (0)	
	703 (0)	ν(C-S) (Met)
725 (2)		
750 (5)	745 (0)	
	766 (5)	Trp
790 sh	788 (0)	
820 (5)	832 (4)	Tyr
852 (4)	860 (4)	Tyr
	888 (0)	
900 (3)	900 sh	ν(C-C) (bracketed 900–965)
930 sh		
940 (6)	945 (2)	
955 sh	965 sh	
	985 (5)	SO_4^{2-}
1004 (7)		Phe
	1009 (8)	

Bovine Serum Albumin	β-Lactoglobulin	Tentative Assignment
1032 (2)	1035 (2)	Phe
1060 (2)		
1080 (2)	1080 (2)	ν(C-N) (bracketed 1080–1162)
	1092 (2)	
1110 (2)	1108 sh	
1127 (4)	1130 (3)	
1160 (2)	1162 (1)	
1180 (3)	1190 (0)	Tyr + Phe
1210 (5)	1212 (1)	Tyr + Phe
	1242 (9)	amide III (bracketed 1242–1262)
1250 (4)		
	1262 sh	
1280 sh	1285 (0)	
1317 (7)	1321 (4)	CH def. (bracketed 1321–1349)
	1325 (?)	
1337 (8)	1349 (6)	
	1365 sh	Trp
1415 (3)	1405 (3)	$(COO)^-$, ν_{sym}
1450 (10)	1455 (10)	CH_2 def.
	1558 (4)	Trp
1660	1660	H_2O

*Frequencies in cm^{-1}; relative intensities in parentheses, with strongest band assigned the value of (10). Abbreviations: Trp, tryptophane; Phe, phenyl; Tyr, tyroisine; Met, methionine. (Reference 33)

All three tyrosines in this protein appear to be "buried" in the interior of the molecule, similar to ribonuclease A (see p.279), when it is dissolved in water. The differences existing between the spectra of cobramine B, recorded between 300 and 1800 cm^{-1}, in the solid and solution may be interpreted in terms of a difference in conformation. In both spectra, a single band is observed in the amide I region (1672 cm^{-1}), suggesting a large fraction of antiparallel β structure. This is supported by the presence of a band at 1235 cm^{-1} (amide III), which is also characteristic of β structure. Some hydrogen-bonded random coil and some α helix probably coexist with β structure, as evidenced by bands at 1254 and 1270 cm^{-1}.

10.6.8 Hemoglobin Derivatives

Numerous physical techniques have been employed to obtain information on the structure and function of hemoglobin, the respiratory protein of red blood cells. Knowledge pertaining to the three-dimensional structure of hemoglobin and its prosthetic group heme (I) has been derived principally from X-ray crystallography, NMR, ESR, and Mossbauer spectroscopy. Only recently has laser Raman spectroscopy been applied to the investigation of this protein (37-41).

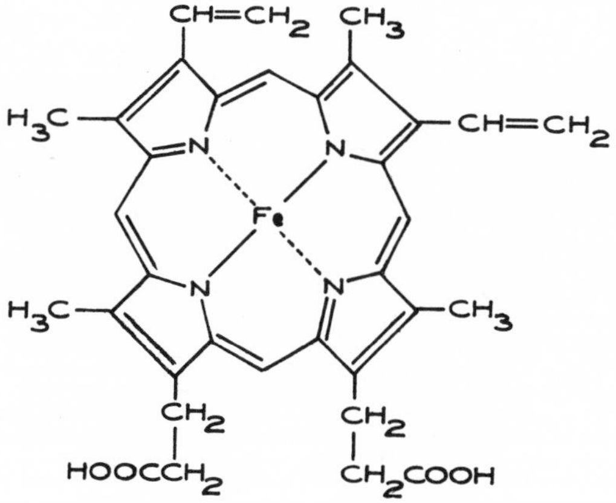

I

Heme is the chromophore of hemoglobin and its accessible electronic transitions are in-plane π-π* transitions. Consequently, the prominent bands in the resonance Raman spectrum (see p.10) should arise from those vibrations that affect the π conjugation in the porphyrin ring. In hemoglobin the heme is attached to the protein only by coordination of an imidazole residue to an axial site on the iron atom (42). The other axial ligand is exogenous. The four heme groups of hemoglobin can be studied by resonance Raman spectroscopy without interference by scattering of the globulin. The reversible and cooperative binding of oxygen by hemoglobin is associated with changes in the structure of the heme group. The principal structural alteration is the movement of the iron atom (43,44) that is delocalized out of the heme plane in deoxyhemoglobin and moves toward the plane of the heme group on ligation in oxyhemoglobin. In addition, a distortion of the entire heme group occurs.

The resonance Raman spectra of oxy- and deoxygenated hemoglobin (488.0 nm Ar^+ excitation) show many differences (37) (Figure 10.10). An important feature is that the frequency shift of the intense band at 1376 cm^{-1} in oxyhemoglobin to 1355 cm^{-1} in deoxyhemoglobin proves whether the hemoglobin under examination is oxygenated or reduced. A band near 1220 cm^{-1}, assigned to the methine C-H in-plane bending vibration in metallooctaethylporphyrins, has been shown to be dependent on the strength of the coordination between porphyrin and metal ion (45). The frequency increase of the 1210 cm^{-1} scattering in deoxyhemoglobin to 1223 cm^{-1} in oxyhemoglobin (Figure 10.10) may arise in the same way. It is probable that some of the other resonance Raman spectral changes observed between oxyhemoglobin and deoxyhemoglobin are associated with the minor structural alterations of the heme group moving from a nearly planar to a more puckered conformation. A red blood cell consists of hemoglobin (30% v/v), the cell membrane, enzymes for the cell metabolism, and other proteins and polysaccharides. Since the non-heme constituents are colorless, they do not contribute to the resonance Raman scattering of a suspension of erythrocytes (38).

It has been found that the moderate strength 1638 cm^{-1} and intense 1589 cm^{-1} bands in the resonance

Raman spectrum of oxyhemoglobin were missing from the spectrum of deoxyhemoglobin (39). Apparently the corresponding vibrational modes "go out of resonance" when hemoglobin is deoxygenated. A notable feature of the resonance Raman spectra of oxy-, carboxy-, and deoxyhemoglobin, and aquo- and azidomethemoglobin is that the frequencies of corresponding bands do not vary by more than 5 cm^{-1} (39). Although changes in either oxidation or spin state of the iron atoms have considerable effects on the electronic spectra (46), the narrow Raman frequency ranges indicate that these changes do not appreciably alter the electron-density distribution on the porphyrin ring.

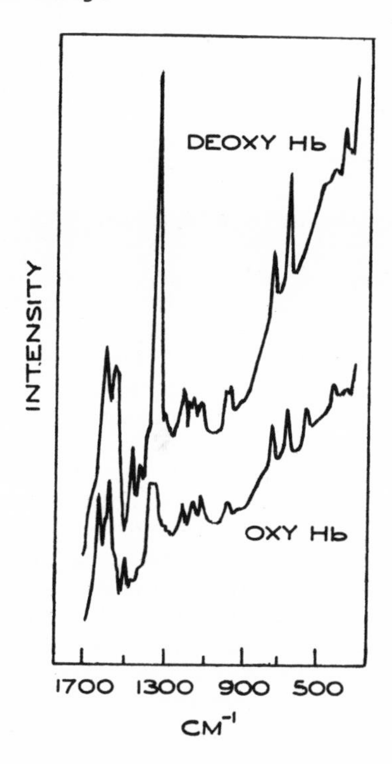

Figure 10.10. Resonance Raman spectrum of 3 x 10^{-5}M oxy- and deoxyhemoglobin in water, pH 65 at 4°C. (The intense line of oxyhemoglobin at 1376 cm^{-1} is truncated for clarity.) (37).

Several of the most prominent bands in the resonance Raman spectra of hemoglobin and cytochrome c, an electron-carrying protein found in mitochondria of all aerobic organisms, display the unusual feature of inverse polarization (40) (see p.30). These bands are active in perpendicular, but absent (hemoglobin) or weak (cytochrome c) in parallel polarization; the depolarization ratio exceeds 100 for the strongest bands. Almost all the remaining bands exhibit depolarization ratios close to the normal depolarized value of 3/4. Spectra obtained for several hemoglobin derivatives and for oxidized cytochrome c show many variations, particularly in relative intensities. However, since the overall polarization pattern remains the same for all the derivatives, it serves to correlate bands from one derivative to another.

An intriguing new development in Raman spectroscopy is the use of a dye laser continuously tunable from 560.0 nm to 640.0 nm. Subsequent to recording the resonance Raman spectra of prepared oxy- and deoxyhemoglobin at several excitation frequencies, an optimum frequency (the visible absorption maximum) was selected to obtain the spectrum of oxyhemoglobin _in vivo_ (10^{-4} M in Fe) (41). Only 20 mW of 563.4 nm laser power was required. The results of this study suggest that it may be feasible to conveniently detect certain diseases of the blood that are related to abnormalities of heme by means of resonance Raman spectroscopy. Generally, slight changes in the environment, altering the nature of substituent groups, and so on, may bring about shifts in the peak visible absorption frequency of a particular compound, causing vibrational modes to "go out of resonance." For this reason, a tunable laser source is necessary to perform a thorough study of substances displaying resonance and preresonance Raman scattering. A very important potential application of the resonance Raman effect is the use of lasers that can be tuned to selectively excite the Raman spectrum of a single biological constituent in a complex living cell in the presence of others excitable at different frequencies.

10.7 CAROTENOIDS

The intense resonance Raman spectra of conjugated polyenes (47-50) and carotenoids (51) in solution had suggested that resonance Raman scattering from pigment molecules in biological samples would be observed without interference from background emission (52) (see p.49). Maximum signal enhancement is gained in carotenoids, the first visible absorption band of which is centered within the frequency range of the Ar^+ laser lines. This effect has been demonstrated in tissue samples of live carrot root and in live tomato fruit, in which the pigment is just one of the constituents (52). Significant frequency differences occur between the hexane solution of lycopene on the one hand and the live tomato fruit and bottled tomato sauce on the other hand. These shifts have been interpreted to arise from solvent effects. From a study of model compounds, inferences might be drawn on the local environment of the pigment in the live tissue. The band widths and shapes for the live samples of carrot root and tomato fruit differ from those for the canned samples.

10.8 CHLOROPHYLLS

Subcellular green bodies called chloroplasts are found in the leaves of all green plants, the color of which is due to chlorophyll. It is believed that chlorophyll exists *in vivo* in a variety of forms, and that these forms represent different states of aggregation of chlorophyll. Aggregation is envisaged to involve coordination of ketone and aldehyde carbonyl oxygen atoms in one chlorophyll molecule with the central magnesium atom of another. Intermolecular aggregation of chlorophylls a and b (II) have been inferred from the measurements of the IR (53,56), UV (57), and NMR (58) spectra of these materials in nonpolar solvents.

Resonance Raman spectra have furnished information pertaining to aggregation *in vivo* (59). No abnormalities are observed in the UV and IR spectra of chlorophylls a and b prepared in the usual way (60), but their resonance Raman spectra reveal the characteristic bands for neoxanthin, a carotenoid present in approximately 0.1% concentration (51). Stringent purification is required to eliminate these bands. The resonance Raman spectrum (476.5 nm Ar^+ irradiation) of an aqueous solution of fragmented chloroplasts extracted from spinach leaves contains bands attributable to both carotenoids and the chlorophylls. Upon decreasing the laser wavelength to below 466.0 nm, the wavelength of the first visible absorption band of neoxanthin, the relative intensities of the carotenoid emissions are significantly reduced while those of the chlorophylls are increased. The chlorophyll a bands are almost completely masked by scattering from chlorophyll b. The spectrum of the latter (10^{-2} M in acetone, nondehydrated) differs from the chlorophyll in the chloroplasts mainly in the 1600-1700 cm^{-1} and 200-600 cm^{-1} regions. This observation may be attributed to difference in the state of aggregation.

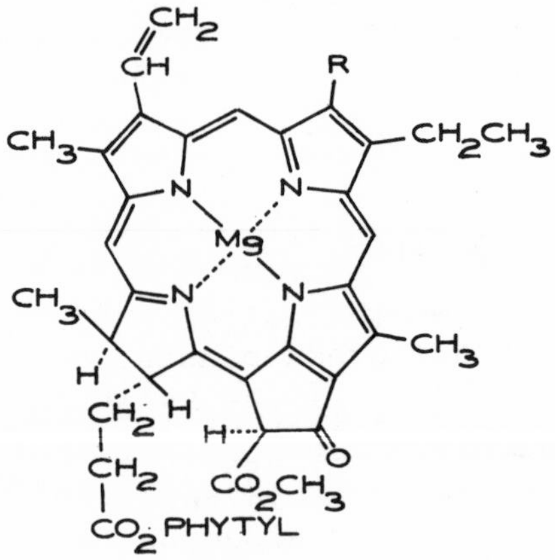

II

CHLOROPHYLL A: $R = CH_3$

CHLOROPHYLL B: $R = CHO$

10.9 NUCLEIC ACIDS AND POLYNUCLEOTIDES

Nucleic acids consist of three components: an organic base or bases, phosphoric acid, and a sugar. The factors responsible both for the secondary structures and the mechanism of enzymatic replication of nucleic acids reside in the nitrogenous bases of the polynucleotide chains. Specific secondary structures are produced by base stacking, or preferential orienting, and base pairing through hydrogen bonds. Vibrational spectroscopy potentially offers very sensitive and detailed descriptions of polynucleotide structures. For example, it has been used to determine the extent of base pairing in RNA's by virtue of the fact that spectra of the components and their possible interaction products are sufficiently different to allow them to be distinguished from each other in mixtures. General methods of characterizing the secondary structures and ascertaining the nature of the specific interactions of nucleic acids in solution are limited. It appears from the results of a few Raman studies in this field that Raman spectroscopy will be a valuable source of structural information.

Intense bands, characteristic of the purine and pyrimidine bases of nucleic acids, are observed which are not appreciably affected by protonation, deuteration, ionization of the phosphate groups, changes in secondary structure, and so forth (Table 10.8) (61). On the other hand, dramatic changes in Raman spectra occur with changes in pH for aqueous solutions of nucleic acid bases that originate from solvent-solute interactions (Figure 10.11) (62).

The intensity ratio of a Raman band attributable to a nucleic acid base residue versus that assignable to the phosphate group sometimes is markedly greater with 488.0 nm Ar^+ excitation when with 632.8 nm He-Ne radiation (63). This fact may be explained by a pre-resonance Raman effect (see p.10). Since nucleic acid bases are characterized by strong absorptions near 260 nm ($\varepsilon \sim 10{,}000$) whereas the phosphate group has no absorption above 180 nm, it is probable that pre-resonance takes place with the bands of about 260 nm.

10.9.1 Adenosine Phosphates

Adenosine phosphates play an important role in biological energy transfer reactions. Vibrational spectroscopy has been employed to gain an insight into the conformational changes associated with pH and with their metal ion complexes (64). Comparison of the Raman spectra of the three adenosine phosphates in pH regions corresponding to the terminal proton ionization reveals that the AMP spectrum is considerably different from both ADP and ATP, while the spectra of the latter two are easily distinguished by the relative intensities, frequencies, and shapes of the bands in the neighborhood of 1125 cm^{-1}. In addition, ADP and ATP have bands at 680 and 710 cm^{-1}, respectively, while AMP has no band in this region.

By studying the pH dependent frequency of particular Raman bands in the spectrum of ATP metal ion complexes in 20-nM aqueous solution, the dissociation of both the secondary phosphate proton and the base proton can be determined (65). For ATP alone, or its complexes with Ca and Mg, the phosphate Raman band changes only at the pK_a for phosphate dissociation, while in the Zn and Mn complexes this same band also shifts at the pK_a for dissociation of adenine, indicating that Zn^{++} or Mn^{++} binding involves both moieties. Quantitative stability constants have been obtained, suggesting weaker binding of Zn and Mn than of Ca and Mg. The fact that Zn and Mn interact more strongly with the base than the biologically more important Ca and Mg tends to support the idea that in living systems it is the metal-phosphate interaction that plays the fundamental role in ATPase mechanisms.

10.9.2 Ribosomal RNA (rRNA)

The Raman spectra of aqueous and D_2O solutions of RNA extracted from ribosomes of E. coli is shown in Figure 10.12 and assignments are given in Table 10.9 (66). The spectrum is not a simple superposition of the spectra of the mononucleotides. The intensity of several bands are affected by an increase in ionic strength, particularly the one at 814 cm^{-1}. This band is attributed to a highly specific structure in the sugar-phosphate linkage of either mono- or polynucleotides. Also, intensity changes between 1450 and 1750 cm^{-1} are interpreted to result from the formation of adenine-uracil and guanine-cytosine base pairs at the expense of unpaired bases.

10.9.3 Transfer RNA (tRNA)

The Raman spectrum of purified formylmethionine tRNA from E. coli has been recorded in aqueous solution using the 514.5 nm Ar^+ line (67). (A much higher background emission was observed for 488.0 nm Ar^+ emission.)

TABLE 10.8. Characteristic Raman and IR Frequencies For Purine and Pyrimidine Bases (61)

Conditions of Observations	ν (cm^{-1})	Intensity Raman	Intensity IR
Uracil Ring			
Solids and all solutions	755-898[a]	s	vw
	795-830	m	s
	1380-1400	s	m-s
All forms except alkaline solutions: Non-deuterated	1230-1280	vs	vw
All forms except alkaline solutions: deuterated	1247-1271	vs	vw
Alkaline solutions only	1010-1035	m	w
	1205-1215	s	w
	1275-1230		
Adenine Ring			
Solids and all solutions	703-733	s	w-m
	1330-1346	vs	m-s
All acidic forms	1403-1421	vs	w
Non-acid forms			
(1) Solids and solutions	1306-1315	s-vs	s
	1450-1490	m-s	m
(2) Solutions only	1378-1391	m-s	--
	1422-1428[b]	w-m	--
Cytosine Ring			
Solids and all solutions	775-795	s-vs	m-s
	1200-1226	s	w-m
	1263-1312	vs	s
	1365-1399	w-m	s
Non-deuterated forms	965-1005	m-s	vw
Acidic solutions	1410-1450	m-s	w
Guanine Ring			
All solution forms	615-680	m	--
	1348-1372	m-s	--
All acidic forms	1260-1295	m-s	--
	1395-1410	s-vs	--
All non-acidic forms	1301-1329	m	--
	1455-1490	vs	--
Alkaline forms only; OH, OD	1323-1345	vs	--
Deuterated forms	1165-1192	w-m	--

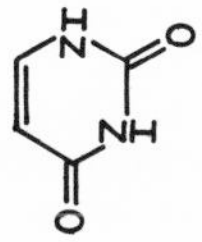

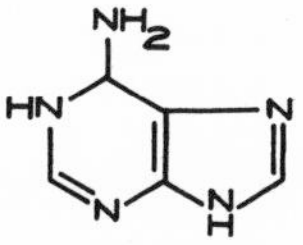

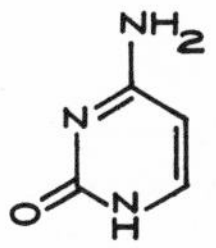

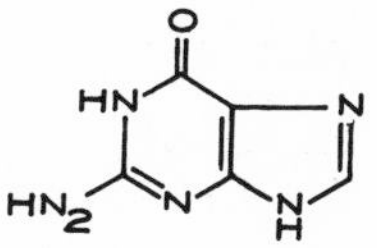

[a] 1,3-dimethyluracil: 687-695

[b] Missing in adenine

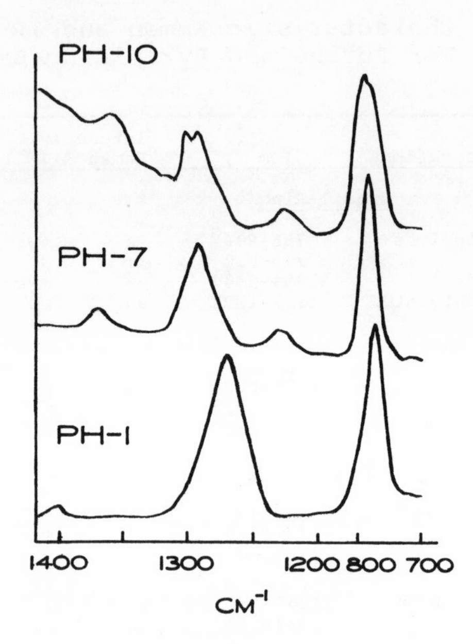

Figure 10.11. Raman spectra of cytosine in aqueous solutions of different pH (62).

This complex molecule is a single polyribonucleotide chain with 77 nucleotide residues, of which 24 are those of guanylic acid, 25 cytidylic acid, 15 adenylic acid, and 8 uridylic acid. The remaining five are residues of odd nucleotides, all differing from one another. The Raman spectrum was compared with the spectra of the homopolymers of the four ribonucleotides: polyriboguanylic acid, poly (G); polyribocytidylic acid, poly (C); polyriboadenylic acid, poly (A); and polyribouridylic acid, poly (U). Since the chemical structure of each nucleotide residue is exactly the same as in tRNA, differences between the Raman spectra of a nucleotide residue in the homopolymer and in tRNA may be attributed to differences in their secondary structures. The Raman spectra of the four homopolymers and tRNA are shown in Figure 10.13. A synthetic spectrum was constructed from these homopolymer spectra by normalizing the Raman line intensities so that the intensity of the symmetric stretching vibration of the PO_2^- group (1000 cm^{-1}) was 1.0.

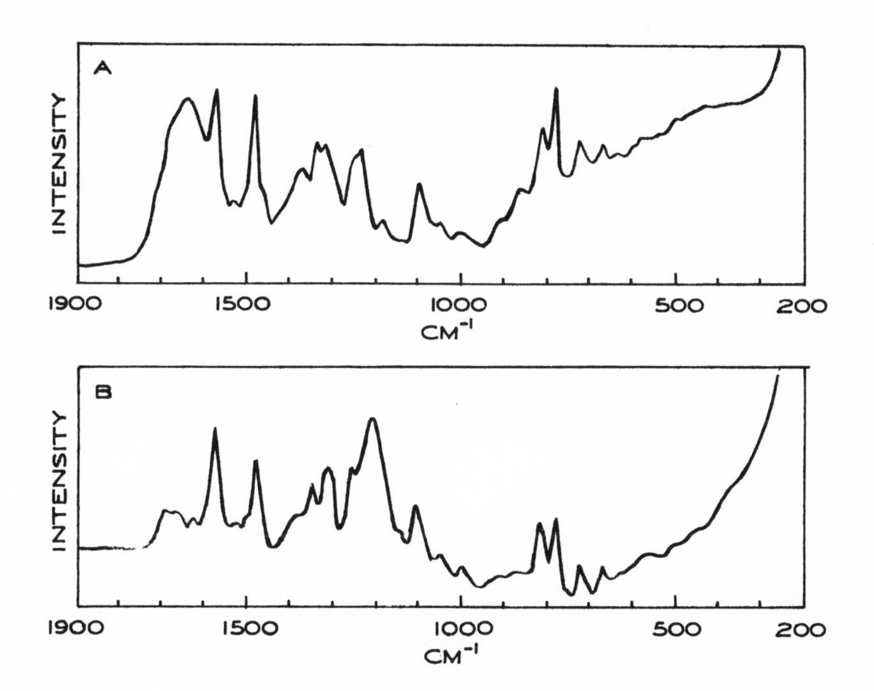

Figure 10.12. Raman spectra of RNA from E. coli in H_2O and D_2O solutions (pH 7, 35 mg/ml). (a) H_2O solution. The strong, broad band at 1640 cm^{-1} is due to H_2O. (b) D_2O solution. The intense broad band at 1208 cm^{-1} is due to D_2O (66). [Reproduced from Biochem. Biophys. Acta]

Every polyribonucleotide has a medium strength PO_2^- Raman line at about 1000 cm^{-1} and it was assumed that its absolute intensity remained constant for any nucleotide residue. These intensities were then multiplied by a factor corresponding to the tRNA composition, that is, 24 for poly (G), and so forth. Comparison of the spectra with that of tRNA revealed several differences which were interpreted as reflecting differences in the average environment of the base residues in the tRNA molecule from those in the homopolymers. The observed lowering of a Raman band frequency indicated a decrease in some bond stretching constants in the base residue, probably caused by hydrogen bonding between bases (G·C and A·U pairing). Furthermore, reduction of the intensity was attributable mainly to the stacking between bases.

TABLE 10.9. Raman Frequencies of Aqueous RNA (66)

H_2O* solution	D_2O* solution	Assignments	
		Nucleotide	Probable origin
435 (0)		Ura, Cyt	Out-of-plane ring deformations; C=O deformations, *etc*....
500 (1)	498 (1)	Gua	
580 (1B)	560 (1B)	Ade, Ura Gua, Cyt	
635 (0)	625 (0)	(Ade), Ura, Cyt	
670 (2)	668 (2)	Gua	Ring stretching
710 (0S)	705 (0S)	Cyt	
725 (3)	718 (3)	Ade	Ring stretching
	755 (0)		
786 (6)	780 (6)	Ura, Cyt	Ring stretching
814 (5)	814 (6)	Phosphate	Symmetric stretching
867 (2)	860 (2B)	Ade, Ura, Gua, Cyt	Ring stretching
918 (1)	915 (1)	Sugar, phosphate	-C-O- stretching
975 (0)		Sugar, phosphate (?)	
	990 (1)	Sugar, phosphate (?)	
1003 (1)		Ade, Ura, Cyt	
1049 (2)	1045 (1)	Sugar, phosphate	-C-O- stretching
	1090 (S)	?	
1100 (5B)	1100 (4)	Phosphate	Symmetric stretching
	1140 (0)	Ade	
1185 (2)	1185 (1B)	Ade, Ura, Gua, Cyt	Ring; external C-N stretching
	1235 (S?)	Ade, Cyt	
1243 (6)		Ura, Cyt	Ring stretching
1255 (5)	1257 (5)	Ade, Cyt	Ring stretching
1300 (4S)		Cyt	Ring stretching
	1310 (7)	Ade, Ura, Cyt	Ring stretching
1320 (7)	1318 (6S)	Ade, Gua	Ring stretching
1340 (7)	1345 (7)	Ura	Ring stretching
	1370 (3B)	Ade, Gua	
1380 (5B)		Ade, Ura, Gua	
	1390 (2S)	Ura	
1422 (2S)		Ade, Gua	Ring stretching; CH deformations
1460 (2S)	1460 (S)	Ura, Cyt	
1484 (10)	1480 (8)	(Ade), Gua	Ring stretching
1510 (S?)	1503 (2S)	Cyt	
1527 (2S)	1526 (3)	(Ade), Cyt	Ring stretching
	1560 (2S)	Ura	
1575 (8)	1578 (10)	Ade, Gua	Ring stretching
1620 (BS)	1622 (3)		Double bond stretching vibrations of paired and unpaired bases; mainly C=O stretching
1650 (BS)			
	1658 (4B)		
1692 (4B)	1688 (4B)		

*Frequencies in cm^{-1}; relative intensities in parentheses, with strongest band assigned the value of (10). The symbols B and S denote broad and shoulder, respectively. Abbreviations: Ade, adenine; Ura, uracil; Gua, guanine; Cyt, cytosine

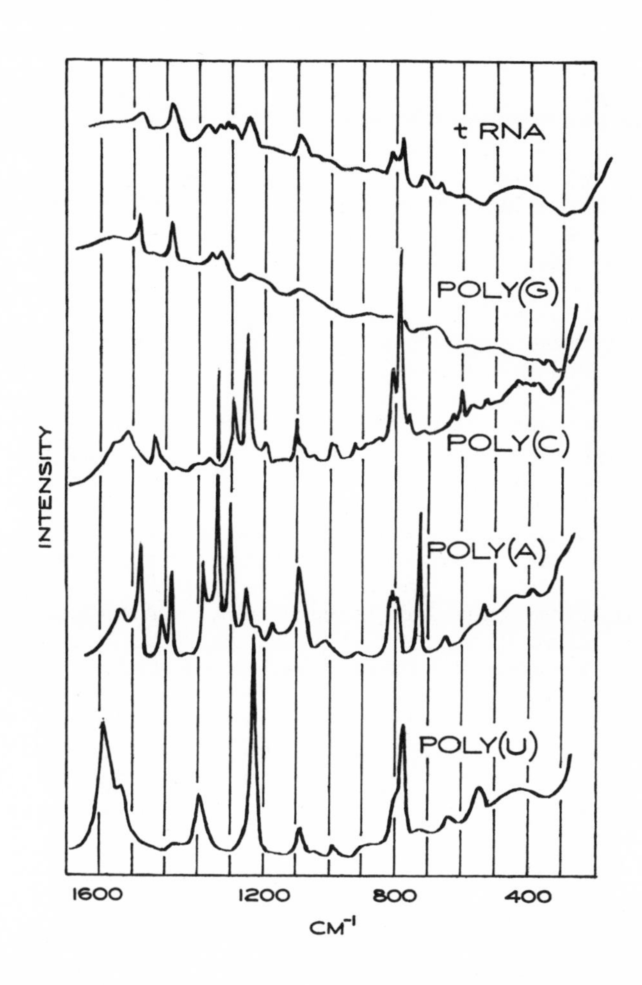

Figure 10.13. Raman spectra of formylmethionine tRNA of E. coli, poly(G), poly(C), poly(A), and poly(U) in aqueous solutions at pH 7.5 at room temperature (67). [Copyright 1971 by the American Association for the Advancement of Science]

10.9.4 RNA Virus R17

The first published Raman spectrum of a native bacterial virus, or phage, appeared in 1973 (68). Small RNA containing viruses consist of one molecule of RNA enclosed by an external coat (capsid) composed of protein subunits (capsomers). In phage R17 the capsomers are arranged on the surface of an icosahedron. The RNA molecule might simply reside in the cavity of the capsid, with few RNA-protein interactions. In this case the secondary structure of RNA in the virus would closely resemble the secondary structure of RNA in solution. Alternatively, the RNA molecule could be situated between adjacent capsomers. In this instance the secondary and tertiary structures of RNA in the virus could be appreciably different from the structure of RNA in solution.

The Raman spectrum of R17 virus contains a large number of Raman bands assignable to the nucleotide residues of RNA and the amino acid residues of protein capsomers (68). Although comparison of Raman spectra of the phage and protein-free RNA suggests many similarities of RNA structure, the average configuration of guanine residues in the phage appears to be very different from that of protein RNA. This observation indicates that guanine plays an important role in RNA-protein interaction. The results of this study suggest that the Raman spectrum of a native virus can reveal many characteristic vibrations of both its protein and nucleic acid components and that Raman spectroscopy is therefore a feasible technique for the study of virus structures in an aqueous environment.

10.9.5 Poly-(adenylic acid) (Poly A)

Poly A, a synthetic polynucleotide, exists mainly in two forms: a double stranded helix below pH 6.0, dissociating on increasing the pH or temperature to an ordered, single stranded conformation (69). The bases are stacked in the "anti" conformation with respect to the sugar base torsion angle, and the turn of the screw axis is right-handed.

The Raman spectrum of Poly A-K salt as a solid (single stranded helix) is shown in Figure 10.14. Similar to other adenine derivatives, the spectrum is dominated by the strong bands of the adenine ring at 725 and 1335 cm^{-1}. Little spectral change is observed on dissolution in water, but heating the solution from 20° to 80° produces an "unstacking" of the chain which is

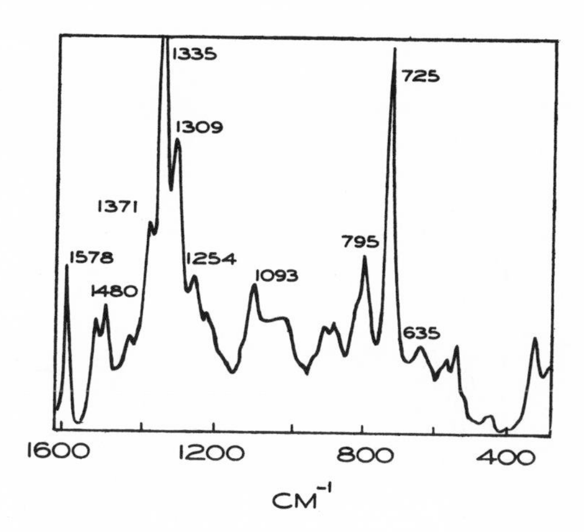

Figure 10.14. Raman spectrum of the potassium salt of poly-(adenylic acid) in the solid state (69).

reflected in a substantial hypochromic change in intensity of a group of bands (Figure 10.15).

10.10 LIPIDS

10.10.1 Relative Intensities of Methylene and Methyl Stretching Vibrations

Information on the packing of hydrocarbon chains in lipids has been obtained from IR frequencies and shapes of methylene rocking vibration bands near 720 cm^{-1} (70). Chain packing types with all hydrocarbon chain planes parallel show a single absorption at 720 cm^{-1}, while those with every second chain perpendicular to the rest display two bands separated of about 10 cm^{-1}. A broad band in this region is observed in disordered states with a hexagonal chain arrangement or liquid chains.

Typical Raman spectra for the liquid and solid states of simple lipids, in which the nonpolar part of the molecules consists of saturated hydrocarbon chains, are shown for tripalmitin in Figure 10.16 (71).

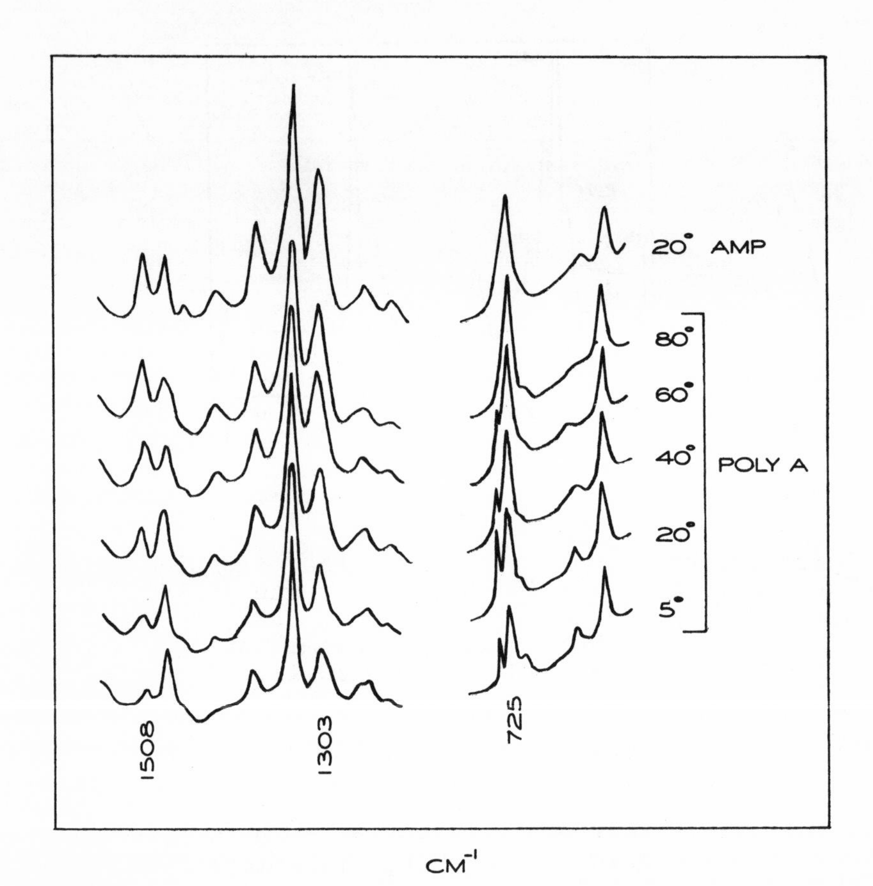

Figure 10.15. Raman spectra of poly-(adenylic acid) as a function of temperature (69).

A common feature of lipids is the striking differences in the 2900 cm^{-1} region between the liquid and solid states. The symmetric stretching vibrations of the CH_2 groups at about 2850 cm^{-1} dominate in the liquid phase, whereas the symmetric CH_3 stretching vibrations near 2800 cm^{-1} are intense in the crystalline state. The large difference in intensities might be related to the fact that all the CH_3 moieties in crystalline phases are located in phases where vibrational interference by lateral interaction may occur between neighboring groups.

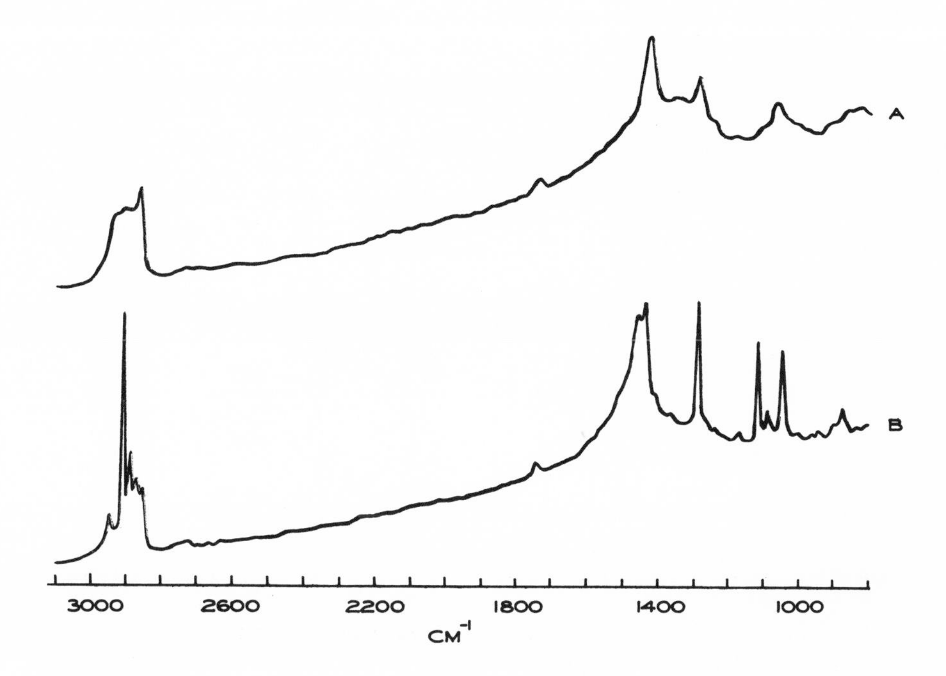

Figure 10.16. Raman spectra of tripalmitin in the β-crystal form (b) and in the liquid state (a) (71).

On the other hand, in the liquid state most of the terminal methyl groups are surrounded by CH_2 groups. Raman bands associated with the most common arrangements of hydrocarbon chains (orthorhombic and triclinic) are summarized in Figure 10.17. Little difference is observed between the Raman spectra of lipids with hexagonal and orthorhombic chain packing, but they can be identified from their IR spectra.

The solid/liquid ratio of plastic fats may be determined from the intensity ratio of the CH_2 stretching vibrations relative to those of the CH_3 group (71). Solid and melted samples of the system are used as references. The Raman spectra recorded on human skin _in vivo_ exhibits two peaks near 2850 and 2890 cm^{-1}, corresponding to lipids with liquid hydrocarbon chains. Raman spectra of psoriatic skin, however, show one band only at about 2860 cm^{-1}, indicating that there are differences in hydrocarbon chain structure or conformation of

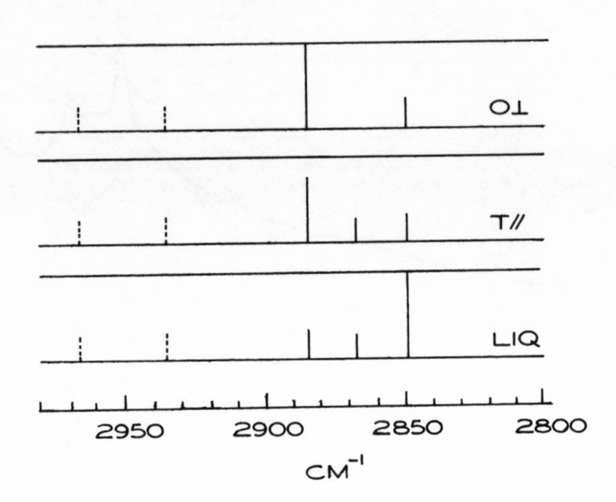

Figure 10.17. Idealized Raman bands of the most common chain-packing arrangements compared to the liquid state. The symmetric vibration bands are indicated by full drawn lines, the asymmetric vibration bands by dotted lines. $O_{\perp}$-orthorhombic (every second chain plane is perpendicular to the rest; $T_{\parallel}$-triclinic (all chain planes parallel) (71).

the lipids in stratum corneum between normal and psoriatic skin (71).

10.10.2 Effect of Cholesterol on Lipid Hydrocarbon Chains

In spite of the fact that cholesterol has long been recognized as an important component of biological membranes, its role remains unclear. Investigations of model systems have suggested that cholesterol acts in different ways, depending on the fatty acid composition of the system. On the one hand, experiments with multilayers of egg yolk lecithin, which contains many unsaturated paraffin chains, indicate that cholesterol considerably rigidizes this lipid (72,73). On the other hand, model systems containing saturated paraffin chains appear to be fluidized by cholesterol at room temperature (74,75). The addition of cholesterol to dipalmitoyl lecithin in water has been found to lower the transition temperature between the gel and liquid crystal phases and to decrease the heat absorbed in the transition (76). It appears from these results that cholesterol controls the phospholipid hydrocarbon chains by disrupting the crystalline chain lattice of the gel phase (fluidization) and by inhibiting the flexing of chains in the liquid crystalline phase (rigidization).

Evidence obtained from Raman spectroscopy indicates that cholesterol affects dipalmitoyl lecithin multilayers by changing the sharp, cooperative gel-liquid crystal transition to a diffuse, noncooperative event by decreasing the interactions between paraffin side chains in the multilayer (77). Consequently, the apparent effect of cholesterol on the multilayers is different above and below the transition temperature. Changes in the Raman spectrum near 1100 cm^{-1}, which can be correlated with structural changes of paraffin chains, were used as a probe to study the effect of cholesterol on the multilayers. The bands in the Raman spectral region 1000-1140 cm^{-1} have been assigned to the skeletal mode of the hydrocarbon chain with a motion such that alternate carbon atoms move in opposite directions along the chain axis (78,79). The Raman spectra of $CH_3(CH_2)_{n-2}CH_3$ solids, where $n > 8$, display two intense bands at 1064 and 1130 cm^{-1}, whose frequencies are independent of n. In addition, a weak band is observed at 1075-1110 cm^{-1} that strongly depends on n. These three bands arise from vibrations of the all-*trans* configuration of the chain. Liquid straight-chain hydrocarbons show a characteristic broad, intense band at 1089 cm^{-1} and weak bands in the narrow ranges 1064-1066 and 1128-1130 cm^{-1}. An intensity decrease of the 1066 and 1130 cm^{-1} bands is attributed to a decrease in the quantity of all-*trans* crystal structures, while the broad band at 1089 cm^{-1} is assigned to structures containing several *gauche* rotations in the melted paraffin (80).

The 990-1190 cm^{-1} region of the Raman spectrum of DL-dipalmitoyl lecithin sonicate (20% by weight in water) at several temperatures is shown in Figure 10.18 (77). As the temperature is changed from 30 to 40°, an abrupt decrease is apparent in the intensities of the Raman bands at 1066 and 1130 cm^{-1} that are due to vibrations of the all-*trans* structure.

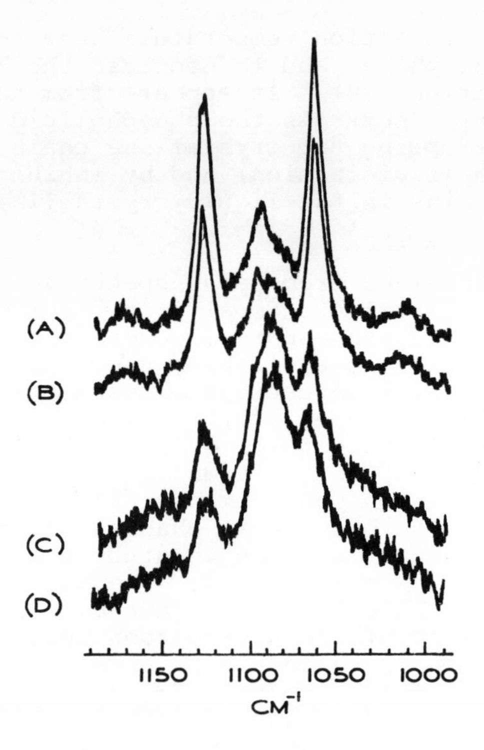

Figure 10.18. Raman spectra of the 1100 cm^{-1} region of 20% (by weight) DL-dipalmitoyl lecithin sonicates in water at (a) 20°C, (b) 30°C, (c) 40°C, and (d) 50°C (77).

Concomitantly, a sharp increase is observed in the intensity of the 1089 cm^{-1} band, assigned to the random liquidlike configurations. The curve obtained by plotting the relative intensities of the 1089 and 1128 cm^{-1} bands against temperature (open triangles) exhibits a sudden change at 38-39°, corresponding to the gel-liquid crystal phase transition (Figure 10.19) (77). This transition involves a change in the palmitate chains from an all-*trans* to a fluid configuration. Since the curve is sharply sigmoidal, this transition is seen for the curve in Figure 10.19 (open circles) that represents a 1:1 mole/mole cholesterol-lecithin sonicate, 20% by weight in water. It shows dramatically that the effect of cholesterol on dipalmitoyl lecithin multi-

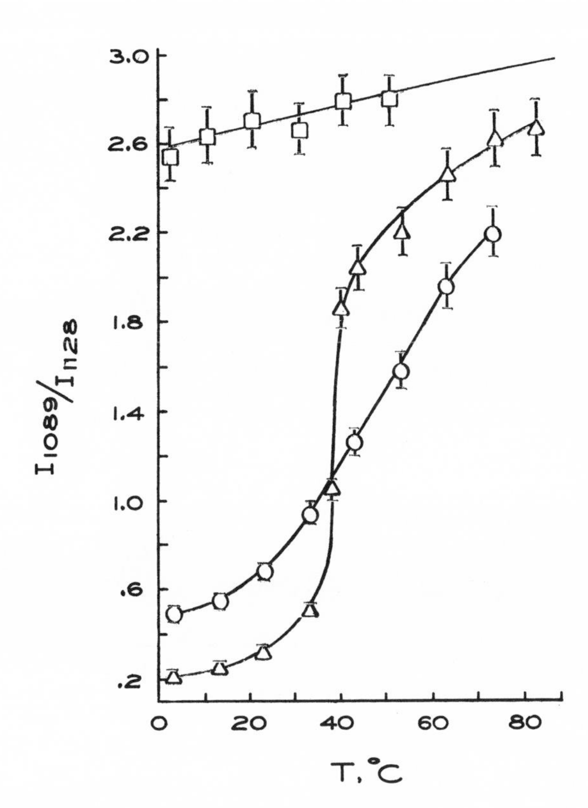

Figure 10.19. Changes with temperature in the ratio of Raman peak heights of DL-dipalmitoyl lecithin at 1089 and 1128 cm^{-1} for (a) 20% (by weight) sonicate in water of dipalmitoyl lecithin Δ, (b) 20% (by weight) sonicate in water of 1:1 cholesterol dipalmitoyl lecithin O, and (c) 10% (by weight) solution of dipalmitoyl lecithin in chloroform □. Similar curves are obtained by comparison of the peak heights at 1089 and 1066 cm^{-1} (77).

layers is to greatly broaden the gel-liquid transition. Thus, multilayers containing cholesterol will display properties more fluid than the pure lipid when measured below the transition temperature, whereas measurements above it will show more rigidity.

Detailed information about the fluidity of hydrocarbon chains in phospholipid-water mixtures also has been obtained by Raman spectroscopy (81).

10.10.3 Erythrocyte Membranes

The Raman spectrum of a suspension of erythrocyte membranes displays four bands in the 1000-1500 cm^{-1} region (1110, 1340, 1420, and 1445 cm^{-1}), attributed mainly to the hydrocarbon chains of the fatty acids (82). The proteins present do not have any strong bands in this spectral region. A single broad band observed at 1110 cm^{-1} indicates that the fatty acid chains are fluid in these membranes.

10.11 VITAMINS

10.11.1 Vitamin B_{12} (Cyanocobalamin)

The electronic spectrum of cyanocobalamin contains bands in the region of 520 and 550 nm that are associated with the corrin ring system (83). The resonance Raman spectra of cyanocobalamin as a solution in 0.1 ammonium sulfate (1.5×10^{-4}M) and as a solid both display an intense band at 1502 cm^{-1} as well as a number of other medium or weak bands (84). The spectra of solutions of dicyanocobalamin and cyanoaquocobinamide are similar above 500 cm^{-1} to those of cyanocobalamin. Since both derivatives possess the intact corrin ring but have different metal-ligand arrangement and side chains, respectively, the spectra of all three substituted corrins can be attributed mainly to the corrin ring.

REFERENCES

1. J. T. Edsall, J. Chem. Phys. 4, 1 (1936).
2. R. A. Russell and W. H. Thompson, Spectrochim. Acta 8, 138 (1956).
3. J. Donahue, Proc. Natl. Acad. Sci. (U.S.) 39, 470 (1953).
4. L. Pauling and R. B. Corey, Natl. Acad. Sci. (U.S.) 37, 256 (1951).

5. T. Miazawa, *Poly-α-Amino Acids*, G. D. Fasman, Ed. (Dekker, New York, 1967).
6. J. L. Koenig and P. L. Sutton, Biopolymers 10, 89 (1971).
7. E. W. Small, B. Fanconi, and W. L. Peticolas, J. Chem. Phys. 52, 4369 (1970).
8. P. L. Sutton and J. L. Koenig, Biopolymers 9, 615 (1970).
9. J. L. Koenig and P. L. Sutton, Biopolymers 8, 167 (1969).
10. J. L. Koenig and P. L. Sutton, Biopolymers 9, 1229 (1970).
11. A. Lewis and H. A. Scheraga, Macromolecules 4, 539 (1971).
12. N.-T. Yu and C. S. Liu, J. Am. Chem. Soc. 94, 5127 (1972).
13. N.-T. Yu and C. S. Liu, J. Am. Chem. Soc. 94, 3250 (1972).
14. B. Fanconi, B. Tomlinson, L. A. Nafie, W. Small, and W. L. Peticolas, J. Chem. Phys. 51, 3993 (1969).
15. R. C. Lord and N.-T. Yu, J. Mol. Biol. 50, 509 (1970).
16. J. L. Koenig, Case Western Reserve Univ. (private communication, 1972).
17. D. F. H. Wallach, J. M. Graham, and A. R. Oseroff, FEBS Letters 7, 330 (1970).
18. M. C. Tobin, Science 161, 68 (1968).
19. R. C. Lord and R. Mendelsohn, J. Am. Chem. Soc. 94, 2133 (1972).
20. C. C. McDonald and D. Phillips, J. Am. Chem. Soc. 91, 1513 (1969).
21. K. Hamaguchi and H. Sakai, J. Biochem. Tokyo 57, 721 (1965).
22. H. Brunner and H. Sussner, Biochem. Biophys. 271, 16 (1972).
23. J. S. Cohen and O. Jardetzky, Proc. Natl. Acad. Sci. (U.S.) 60, 92 (1968).
24. C. C. McDonald and W. D. Phillips, J. Am. Chem. Soc. 91, 1513 (1969).
25. D. F. Waugh, Advan. Protein Chem. 9, 326 (1954).
26. V. du Vigneaud, R. H. Sifford, and R. R. Sealock, J. Biol. Chem. 102, 521 (1933).
27. N.-T. Yu, C. S. Liu, and D. C. O'Shea, J. Mol. Biol. 70, 117 (1972).
28. M. Schiffer and A. B. Edmundson, Biophys. J. 10, 293 (1970).

29. W. B. Gratzer, G. H. Beaven, H. W. E. Rattle, and E. M. Bradbury, Eur. J. Biochem. 3, 276 (1968).
30. W. B. Gratzer, E. Bailey, and G. H. Beaven, Biochem. Biophys. Res. Comm. 28, 914 (1967).
31. A. G. Walton, M. J. Deveney, and J. L. Koenig, Calc. Tiss. Res., 162 (1970).
32. W. B. Rippon, J. L. Koenig, J. Agr. Food Chem. 19, 692 (1971).
33. A. M. Belloco, R. C. Lord, and R. Mendelsohn, Biochem. Biophys. Acta 257, 280 (1972).
34. E. Ellenbogen, J. Am. Chem. Soc. 77, 6634 (1955).
35. N.-T. Yu, B. H. Jo, and C. S. Liu, J. Am. Chem. Soc. 94, 7572 (1972).
36. N.-T. Yu, B. H. Jo, and D. C. O'Shea, Arch. Biochem. Biophys. (in press).
37. H. Brunner, A. Mayer, and H. Sussner, J. Mol. Biol. 70, 153 (1972).
38. H. Brunner and H. Sussner, Biochem. Biophys. Acta B10, 20 (1973).
39. T. C. Strekas and T. G. Spiro, Biochim. Biophys. 263, 830 (1972).
40. T. G. Spiro and T. C. Strekas, Proc. Natl. Acad. Sci. (U.S.) 69, 2622 (1972).
41. P. R. Reed, Spex Instr. Co., Metuchen, N. J. (private communication, 1972).
42. M. F. Perutz, J. M. Cox, and L. C. G. Goaman, Nature 219, 137 (1968).
43. M. F. Perutz, Nature 228, 726 (1970).
44. R. Huber, O. Epp, and H. Formanek, J. Mol. Biol. 52, 349 (1970).
45. H. Ogoshi and Z. Yoshida, Bull. Chem. Soc. Japan 44, 1722 (1967).
46. D. W. Smith and R. J. P. Williams, *Structure and Bonding*, (Springer-Verlag, Berlin, 1970), Vol. 7, p.l.
47. J. Behringer and J. Brandmuller, Ann. Physik. 4, 234 (1959).
48. T. M. Ivanov, L. A. Yanovskaya, and P. P. Shorygin, Opt. i Spektroskopiya 18, 206 (1965).
49. P. P. Shorygin and T. M. Ivanov. Dokl. Akad. Nauk. SSSR, 150, 533 (1963).
50. P. P. Shorygin and T. M. Ivanova, Optik. i Spektroskopiya 15, 176 (1963).
51. L. Rimai, R. G. Kilponen, and D. J. Gill, J. Am. Chem. Soc. 92, 3284 (1970).
52. D. Gill, R. G. Kilponen, and L. Rimai, Nature 227, 743 (1970).

53. L. J. Boucher, H. H. Strain, and J. J. Katz, J. Am. Chem. Soc. 88, 1341 (1966).
54. J. J. Katz, G. I. Kloss, F. C. Pennington, M. R. Thomas, and H. H. Strain, J. Am. Chem. Soc. 85, 3801 (1963).
55. K. Ballschmiter, T. M. Cotton, H. H. Strain, and J. J. Katz, Biochem. Biophys. Acta 180, 347 (1969).
56. A. Rosilio and J. P. Leicknam, Revue of G.A.M.S. 6, 132 (1970).
57. A. F. H. Anderson and M. Calvin, Arch. Biochem. Biophys. 107, 251 (1964).
58. G. I. Gloss, J. J. Katz, F. C. Pennington, M. R. Thomas, and H. H. Strain, J. Am. Chem. Soc. 85, 3809 (1963).
59. M. M. Lutz, C.R.H. Acad. Sci. Paris, Ser. B 275, 497 (1972).
60. R. B. Park and J. Biggins, Science 144, 1009 (1964).
61. R. C. Lord and G. J. Thomas, Jr., Spectrochim. Acta 23A, 2551 (1967).
62. G. Arie, E. DaSilva, G. Dumas, H. Rozansza, and C. Sebenne, Biochimie 53, 1041 (1971).
63. M. Tsuboi, S. Takahashi, S. Muraishi, and T. Kajiura, Bull. Soc. Japan 44, 2921 (1971).
64. L. Rimai, T. Cole, J. L. Parsons, J. T. Hickmott Jr., and E. B. Carew, Biophys. J. 9, 320 (1969).
65. L. Rimai and M. E. H. Heyde, Biochem. Biophys. Res. Comm. 41, 313 (1970).
66. G. J. Thomas, Jr., Biochem. Biophys. Acta 213, 417 (1970).
67. M. Tsuboi, S. Takahashi, S. Muraishi, T. Kajiura, and S. Nishimura, Science 174, 1142 (1971).
68. K. A. Hartman, N. Clayton, and G. J. Thomas, Jr., Biochem. Biophys. Res. Comm. 50, 942 (1973).
69. N. N. Aylward and J. L. Koenig, Macromolecules 3, 590 (1970).
70. D. Chapman, *The Structure of Lipids* (Methuen and Co., Ltd., 1964).
71. K. Larson, Chem. Phys. Lipids (in press).
72. A. Finkelstein and A. Cass, Nature 216, 717 (1967).
73. J. C. Hsia, H. Schneider, and I.C.P. Smith, Chem. Phys. Lipids 4, 238 (1970).
74. D. O. Shah and J. H. Schulman, J. Lipid Res. 8(3), 215 (1967).
75. P. Joos, Chem. Phys. Lipids 4, 162 (1970).
76. B. D. Ladbrooke, R. M. Williams, and D. Chapman, Biochim. Biophys. Acta 150, 333 (1968).

77. J. L. Lippert and W. L. Peticolas, Proc. Nat. Acad. Sci. (U.S.) 68, 1572 (1971).
78. M. Tasumi and T. Shimanouchi, J. Mol. Spectros. 9, 261 (1962).
79. R. G. Snyder and J. H. Schachtscheider, Spectochim. Acta 19, 85 (1963).
80. R. G. Snyder, J. Chem. Phys. 47, 1316 (1967).
81. B. J. Bulkin and N. Krishnamachari, J. Am. Chem. Soc. 94, 1109 (1972).
82. B. L. Bulkin, Biochim. Biophys. Acta 274, 649 (1972).
83. A. J. Thompson, J. Am. Chem. Soc. 91, 2780, (1969).
84. W. O. George and R. Mendelsohn, Appl. Spectrosc. 27, 390 (1973).

Chapter 11

REMOTE RAMAN SPECTROSCOPY IN POLLUTION STUDIES

11.1 INTRODUCTION

Human activity has polluted the air with biologically harmful substances and the world-wide dispersion of air pollution presents a growing threat to man's health and land environment. Gaseous air pollutants cause damage to vegetation, irritatión to eyes and other toxic effects, and possible modification of the global climate. Early notice of the effects of air pollution was due to smoke and smog, but it is now realized that many of the damaging effects result from invisible gaseous constituents. Only within the last few years have advances in optical instruments permitted atmospheric gases to be detected and measured remotely from the ground, from aircraft, high flying balloons, and satellites (1).

Laser technology has brought about a revolution in the detection and measurement of air and water pollutants. The low divergence of laser beams and their ability to transfer large amounts of electromagnetic energy over great distances suggest a tremendous potential for the measurement of the concentration of pollutants from a remote location. Considerable effort has been expended on three classes of laser-based systems: (1) Raman scattering, (2) resonance Raman scattering (see p.10), and (3) resonance absorption (2), but only Raman scattering instrumentation has been successfully tested in the field.

11.2 DETECTION AND MEASUREMENT OF AIR POLLUTANTS

11.2.1 Remote Raman Instrumentation

The remote Raman technique is a logical extension of LIDAR (3), which employs a single-ended, remote, range resolved measurement of light backscattered from particles to give a three-dimensional plot of their concentration in the atomosphere without regard for chemical identification or individual species concentration. Operation of the Raman method is based on a radarlike system that uses light instead of radio waves. The schematic for a field-tested Raman system is shown in Figure 11.1 (4). A beam from a high-power UV pulsed laser (frequency doubled ruby laser, 347.2 nm) is sent to a remote sample. A Raman shift occurs in the backscattered light, resulting in spectral bands characteristic of the molecules intercepting the laser radiation. The backscattered light is collected by a large and efficient Cassegrainian telescope, whose field of view is collinear to that of the laser, and focused into the slit of a polychromator that simultaneously sorts several Raman emissions into discrete frequencies. The Raman emissions are simultaneously sensed by a bank of photomultiplier tubes, digitally processed, and the data displayed in a usable format. High dynode gain photomultipliers are used in conjunction with a pulse change digitizer to count the number of photoelectrons in each pulse.

The returning photons are counted during a single pulse, and then a measurement is made of the background radiation coming from the same portion of the field of view for the same length of time. The background photon count is subtracted from the original signal and the difference is stored in the system. After integrating over several pulses the data are displayed and ratioed against a reference channel which views the N_2 return from the same portion of space at the same time that the other channels are operating. Hence, atmospheric transmission and the relatively large pulse-to-pulse variations are essentially eliminated. Final data are obtained by signal averaging over an observation time of less than one minute. The quantity of a particular compound in the atmosphere is determined by the corrected intensity of the returning radiation. The distance from the laser to the compound is ascertained by sending out a few of the extremely short pulses, which are essentially slices of light chopped into approximately 10-m sections, and observing the time of return.

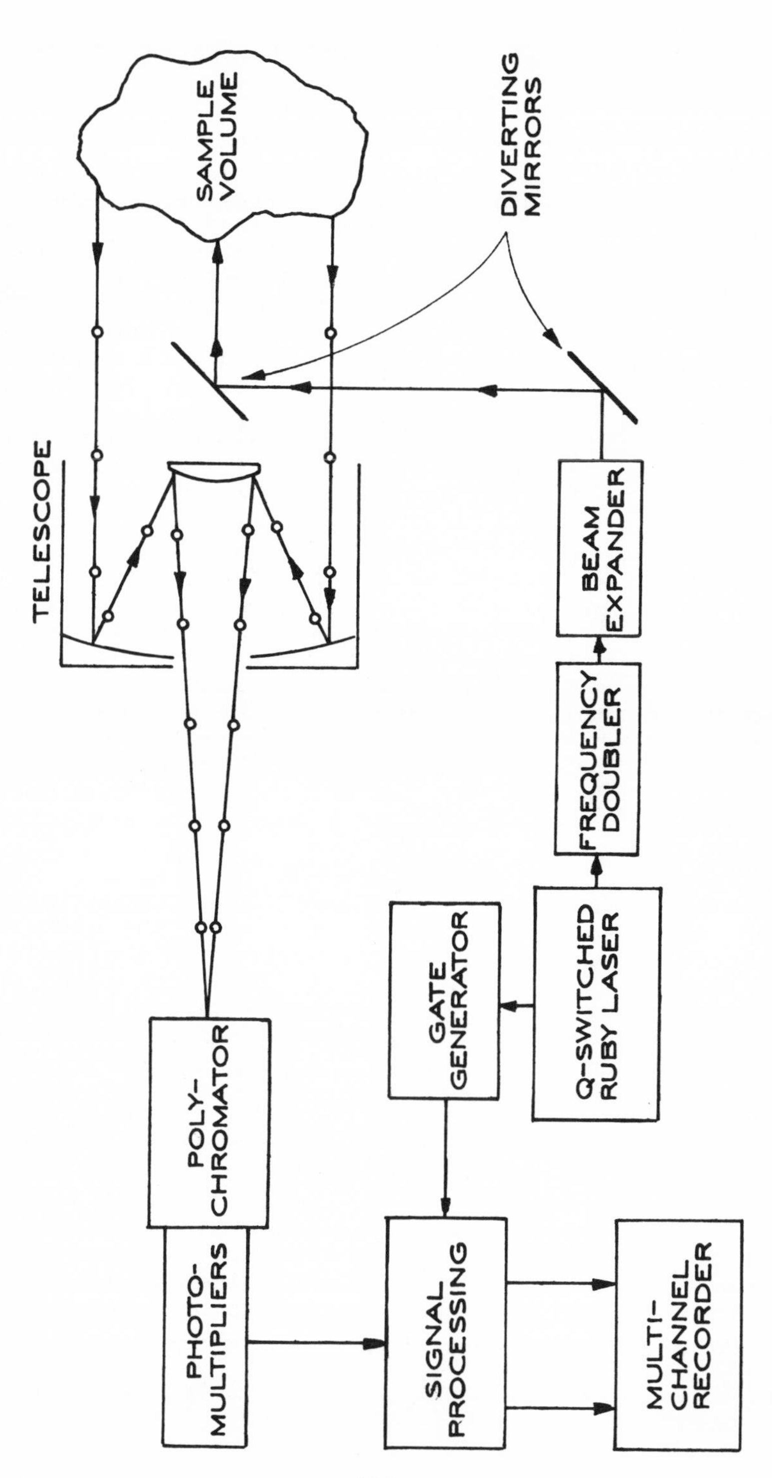

Figure 11.1. Schematic of a remote Raman system (4).

Pointing the laser in various directions then gives the position of the pollutant. The description of a sampled volume appears in Figure 11.2 (5).

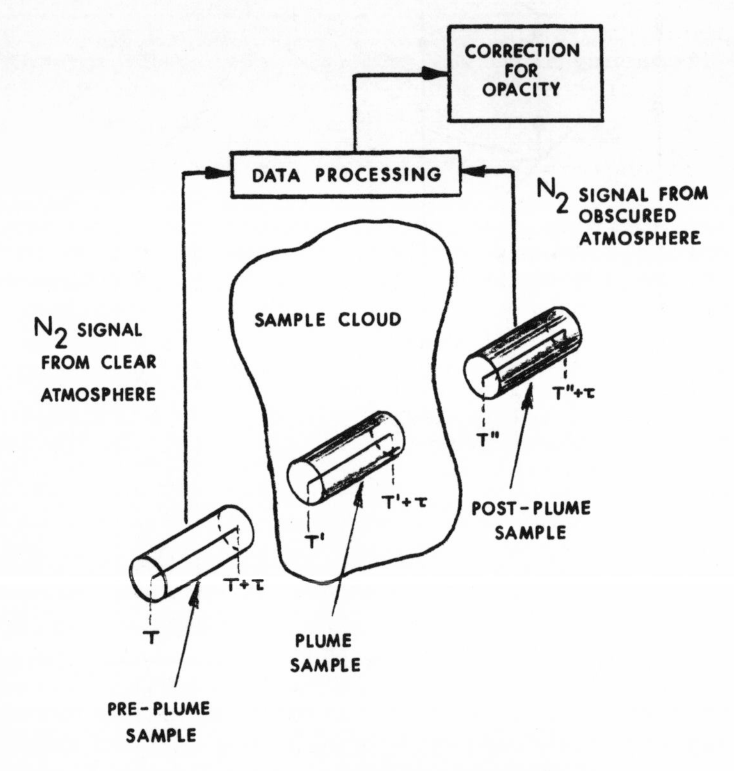

Figure 11.2. Definition of sampled volume for a remote Raman spectrometer (5).

11.2.2 Comparison of Remote Methods

Correlation Spectroscopy

Correlation spectroscopy has been applied in remote measurements of air pollutants. Using this technique the spectrum of the sample is obtained within a suitable frequency interval and then compared (correlated) with that of the pollutant being measured (6). These devices are divided into two categories, depending upon whether or not the spectrum of the incident light is dispersed before the comparison is performed.

UV dispersion instruments (7) are specific, small, and portable, but possess a major disadvantage in that only a small number of pollutants have a UV spectrum. In addition, they are unable to monitor more than one or two pollutants at a time. The typical IR dispersion instrument is energy limited. In order to increase resolution, the slits must be narrowed or the radiation spread widened. Thus, less energy is available to the detector and the signal-to-noise (S/N) ratio is lower. The energy restriction can be improved somewhat by opening the slits, but this results in a loss of resolution. In generating a spectrum the detector measures the intensity of each segment of the radiation and only a limited amount of time can be spent on an individual measurement if a full spectrum is presented within a reasonable length of time. However, the S/N ratio is directly proportional to the measuring time. Hence, the conventional dispersion spectrometer is limited in response speed, resolution, sensitivity, and S/N ratio, and any one characteristic can be materially improved only at the expense of others. On the other hand, in interferometric spectroscopy, various performance characteristics can be improved simultaneously. Of the various IR instruments presently employed in remote measurements, the highest optical efficiency is obtained interferometrically with the use of Fourier transform spectroscopy (8). A commercially available interferometric IR system (9) can scan the 200-4000 cm^{-1} range and is capable of identifying and quantifying any gaseous pollutants whose spectrum is stored in the computer memory in the form of an interferogram. Long-path monitoring of materials in urban areas can be accomplished with a high temperature, broad-band IR source over distances as great as 6000 m. A separate light source is not necessary to monitor exhaust gases from industrial facilities, because the self-emission of the heated pollutants can be used. The IR Fourier transform system is one or two orders of magnitude more sensitive than remote Raman instrumentation and therefore it is able to detect ambient (dispersed) pollutants. On the other hand the lack of range resolution renders the former unsuitable for pollution source location and mapping. Both techniques are sophisticated and expensive.

Resonance Methods

The resonance absorption technique measures the total amount of pollutants in the light path without depth resolution (1). The method requires minimal laser power, is relatively simple, and has high sensitivity. A remote detector or reflective target is necessary to receive the transmitted beam. Resonance Raman spectroscopy offers a considerable enhancement in sensitivity over the ordinary Raman effect but it suffers from the same major disadvantage as the resonance absorption method, that is, adequate tunable lasers are not yet available.

Remote Raman Spectroscopy

The remote Raman method is sufficiently sensitive (about 10 ppm, or about 10 mg/m^3) to measure "emitted" pollution but cannot detect it when dispersed in the atmosphere. This technique has an important advantage over other approaches: range resolution, which makes it not only possible to determine the distance of the measured volume from the instrument, but also the path length through the observed sample provided that total absorption does not occur over the sample path. In contrast, remote IR cannot differentiate between 1 mg/m^3 and 0.1 $mg/10\ m^3$.

Changes in concentration cause variations in peak height, but not peak width of a Raman band. Hence, high moisture content of the air and naturally occurring atmospheric constituents do not interfere with the detection and analysis of trace contaminants. (Since detectability is directly related to visibility, the effectiveness of a Raman system is reduced in rain and fog.) On the other hand, problems concerned with specificity are encountered when remote IR is used in the presence of an unknown background matrix. The remote Raman method is insensitive to moderate temperature fluctuations in the measured volume and essentially is unaffected by variations of ambient pressure and/or density. Prototype remote Raman instrumentation is very expensive and quite large (the unit is operated from a truck trailer), but a second generation system has improved upon these limitations (10).

The difficulty of calibrating remote Raman instruments in the field has been overcome by using trace atmospheric constituents normally present at nearly constant concentrations (11), for example, $^{14}N^{15}N$--2889 ppm at 2292 cm^{-1}, $^{12}C^{16}O_2$--306 ppm at 1388 cm^{-1}, $N^{15}N^{15}$--10.1 ppm at 2302 cm^{-1}.

11.2.3 Experimental

Raman frequencies for some gaseous species of current interest in pollution studies are listed in Table 11.1. Since the remote Raman technique was first proposed (11), a controversy has surrounded the adequacy of its sensitivity for pollution studies (11-16). A recent study reported data collected under field conditions in different geographical locations over a wide range of climatic conditions (10) (Table 11.2). Daylight spectra for SO_2 (30 ppm) and kerosine (1.7 ppm) dispersed in controlled amounts to create plumes of known concentration, are presented in Figure 11.3 (10). An instrument presently in the development stage will have approximately a tenfold increase in sensitivity (17).

Low molecular weight hydrocarbons, which photochemically react with oxygen in the air to form smog, can be determined in as little as 2 ppm concentration at a distance of approximately 400 m from the source (4). The spectral resolution at the analytical wavelength of about 2900 cm^{-1} is insufficient to separate the CH, CH_2, and CH_3 stretching vibrations. Consequently, one band only is observed for the total C-H stretches. Characterization of the hydrocarbon is accomplished by means of the bands appearing between 800 and 1400 cm^{-1}.

In some instances a fortuitous frequency coincidence allows the resonance Raman effect to operate.

TABLE 11.1. Raman Frequencies for Some Gaseous Species of Interest in Pollution Studies

Molecule	ν (cm^{-1})	Molecule	ν (cm^{-1})
CO_2	668	NO^+	2248
O_3	710	N_2	2273
SF_6	775	NO^+	2277
NH_3	950	N_2	2302
Aliphatics	987	NO^+	2305
Aromatics	992	N_2	2331
O_3	1043	CO	2349
SO_2	1151	H_2S	2611
CO_2	1242	O_3	2800
CO_2	1265	Aliphatics	2857
N_2O	1285	CH_4	2914
CO_2	1286	O_3	3050
NO_2	1320	Primary Amines	3189
CO_2	1388	Primary Amines	3256
CO_2	1409	NH_3	3331
CO_2	1430	Primary Amines	3343
CO_2	1528	CO_2	3609
O_2	1556	H_2O	3652
O_3	1740	CO_2	3716
NO	1876	NO^+	4422
O_3	2105	N_2	4459
CO	2145	NO^+	4478
N_2^+	2175	N_2	4517
NO^+	2221	NO^+	4534
N_2O	2223	N_2	4575
N_2	2244	NO^+	4590
		N_2	4633

For example, when a doubled ruby laser is employed, the detection limit for phosphate insecticides (POC group) is less than 0.05 ppm. The frequency of a CO laser falls under the IR absorption band of NO, enabling the detection of very small quantities of this atmospheric pollutant.

In summary, the remote Raman technique appears to offer a unique capability for remote monitoring of a wide range of air pollutants. The applications include large area quantitative monitoring, cloud mapping, tracking of pollutant clouds to their sources, and stack monitoring. Present systems are large, very

TABLE 11.2. Typical Detection Limits of a Remote Raman Spectrometer (Range: 250 Meters--Sample Path 10 Meters) (10)

Compound	ppm
Nitrogen	46
Sulfur hexafluoride	6
Sulfur dioxide	4
Nitrous oxide	17
Nitric oxide	585
Ammonia	11
Water (vapor)	13
Benzene (aerosol)	2
Kerosine (aerosol)	1
Methanol (aerosol)	16
Nitric acid (aerosol)	3
Organophosphates (aerosol)	4

costly, and extremely complex, but it is likely that the technology will be advanced to the point where the technique becomes more practical.

11.2.4 Potential Applications

There are several interesting areas where remote Raman spectroscopy probably will find application (17).

1. Thermal Pollution. Results of a study of flame spectra by the remote Raman method (4) indicate that sensing the temperature of jet engine plumes, stack flumes, and so forth, can be accomplished by modifying existing instrumentation. At present, jet engines are tuned to obtain the maximum thrust for the minimum temperature, with little regard for minimizing pollutant output. Preliminary evidence indicates that emitted particulates do not increase the background signal.

2. Tracer Techniques. Remote Raman systems should be capable of obtaining climatic information by micrometeorological measurements. As an example, the direction of the wind and microturbulances may be determined by adding an SF_6 tracer to the atmosphere. Sulfur hexafluoride has many of the characteristics of an ideal tracer, but its detectability limit is only 6 ppm.

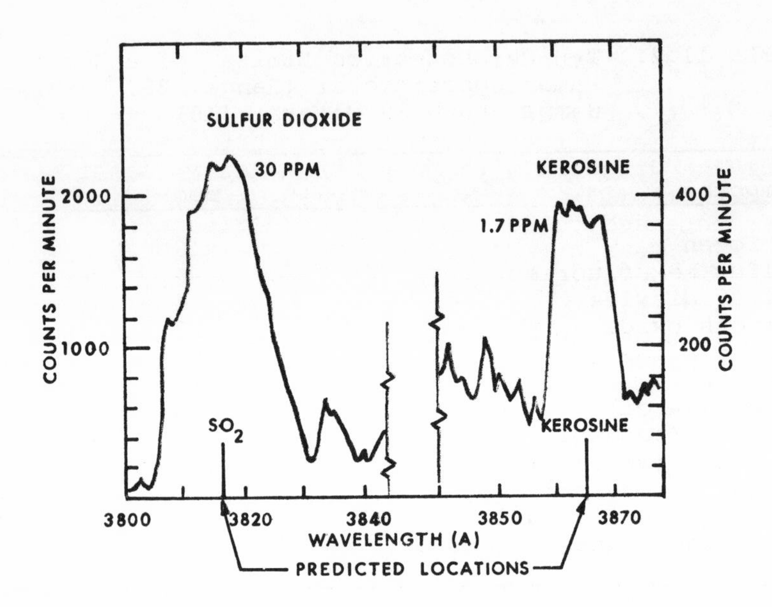

Figure 11.3. Remote Raman returns for SO_2 and kerosine air pollution at 200 m (10).

The use of fluorescent tracers with the same detector systems offers definite advantages.

3. Measurement of HCl. Tons of HCl, originating from rocket fuels containing perchlorate oxidizers, are dumped into the atmosphere as a result of rocket launches. Most of the acid presumably remains in the air, but the possibility exists that some of it falls to earth. An airborne remote Raman system could measure the atmospheric HCl concentration at different heights as well as monitoring the rocket plume.

11.3 DETECTION AND DETERMINATION OF WATER POLLUTANTS

No viable system presently is available for direct or remote Raman sensing and measuring water pollutants under field conditions. Preliminary experiments, however, indicate that Raman spectroscopy may find utility in this area.

11.3.1 Standard Raman Method

The first reported study of water pollutants by means of the Raman effect demonstrated that 50-ppm benzene could be detected with only a 5 mW He-Ne laser (632.8 nm) (18). Subsequently the sensitivity limit was decreased to 23 ppm using a multipass cell and 30 mW He-Ne excitation (19). Under these conditions, 22-ppm CS_2 was detectable (657 cm^{-1} band). The ultimate sensitivity of a low cost, experimental Raman system was shown to be 1000-ppm benzene in water and it was predicted that a one watt Ar^+ laser, coupled with other improvements, could yield detectivities of about 1 ppm (19).

Detection limits for some ionic water pollutants in single-pass cells with commercial instrumentation (20), along with the desired limits of detectability (21), are listed in Table 11.3.

TABLE 11.3. Detection Limits for Some Water Pollutants

Anion	Desired Level (ppm)	Detected by Raman (ppm)	Observed Raman Frequency (cm^{-1})
NO_3^-	10 ± 0.5	25	1049
PO_4^{3-}	20 ± 0.5	50	989
SO_4^{2-}	100 ± 2	50	982
CO_3^{2-}		75	1063

11.3.2 Remote Method

Qualitative and quantitative analysis of trace species in water has been performed using a prototype remote Raman instrument (22). An interesting feature of this work is the conspicuous effect of salinity on the Raman spectrum of water between 3000 and 3750 cm^{-1} (Figure 11.4).

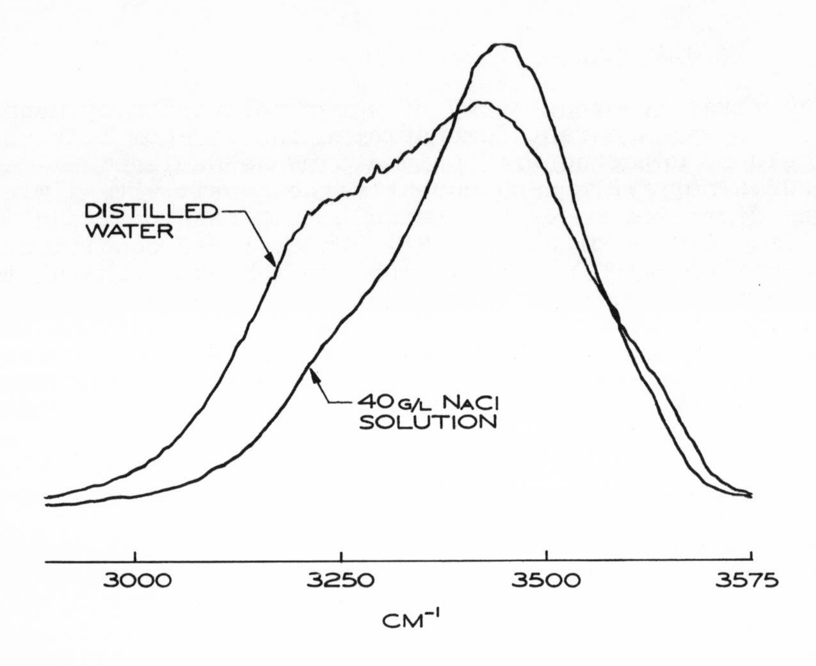

Figure 11.4. Effect of salinity on the Raman spectrum of water (10).

Laboratory measurements performed on KH_2PO_4, KNO_3, and $(NH_4)_2SO_4$ demonstrated detectability limits of about 5, 20, and 30 ppm, respectively.

A very interesting application of remote Raman spectroscopy is concerned with the examination of algae, chlorophyll-containing plants. Chlorophyll *in vitro* fluoresces strongly when exposed to red radiation. On the other hand, *in vivo* chlorophyll, which is associated with a complex biochemical system evolved to convert solar energy to carbohydrates, fluoresces only weakly. At high illumination intensities, however, this system saturates. Observations with a Q-switched laser thus are proportional to total chlorophyll, while the same measurement with the Q switch off gives only the chlorophyll not associated with living systems. This marked difference in fluorescence has been employed to distinguish between live and dead algae in regions containing unusually high algae concentrations (23).

REFERENCES

1. "Remote Measurement of Pollution," NTIS, U.S. Dept. of Commerce No. N72-18324, August 1971
2. H. Kildal and R. L. Beyer, U.S. Dept. of Commerce No. AD730 770 (1970).
3. P. M. Hamilton, Int'l. J. Air and Water Pollution 19, 437 (1966).
4. T. Hirschfeld, S. Klainer, and R. Burton, Proc. Electro Optical Systems Design Symp., New York, Sept. 1969, p.418.
5. H. Tannenbaum, D. Tannenbaum, H. DeLong, E. R. Schildkraut, S. Klainer, and T. Hirschfeld, Paper presented at Sixty-fifth Annual Meeting Air Pollution Control Association, Miami Beach, Fla., June 1972.
6. A. R. Barringer, Am. Astronautical Soc. Sci. and Tech., Ser. 4 (1965).
7. Barringer Research Ltd., Toronto, Ontario, Canada.
8. S. K. Freeman, in *Ancillary Techniques of Gas Chromatography*, L. S. Ettre and W. H. McFadden (Eds.) (Wiley-Interscience, New York, 1969).
9. Digilab Division of Block Eng. Inc., Cambridge, Mass.
10. T. Hirschfeld, E. R. Schildkraut, H. Tannenbaum, and D. Tannenbaum, Appl. Spec. Lett. 22, 38 (1973).
11. T. Hirschfeld, Block Instr. Co. (private communication, 1972).
12. J. Cooney, in *Proc. of Symp. on Electromag. Sensing of the Earth from Satellites*, R. Zirkind (Ed.) (Polytechnic Press, New York, 1965) pp.1-10.
13. T. Hirschfeld and S. Klainer, in Proc. Symp. Laser Appl. Geosciences, Huntington Beach, July 1969, p.69.
14. D. Leonard, NAPCA Report, NTIS Report No. PB198204, December 1970.
15. T. Hirschfeld, S. Klainer, and R. Burton, Proc. Electro Optical Syst. Design Symp., New York, September, 1969, p.418.
16. H. Kildal and R. L. Byer, Proc. IEEE 59, 1644 (1971).
17. E. R. Schildkraut and T. Hirschfeld, Paper presented at the Eastern Analytical Symposium, Atlantic City, N.J., Nov. 1972.
18. E. B. Bradley and C. A. Frenzel, Water Res. 4, 125 (1970).
19. E. B. Bradley and C. A. Frenzel, NTIS, U.S. Dept. Commerce Document PB 208 029 (1971).

20. S. F. Baldwin and C. W. Brown, Water Res. 6, 1601 (1972).
21. R. S. Green, "Analysis Instrumentation," in *Proc. of Twelfth Annual Symp., Houston, 1966* (Plenum Press, New York, 1967), Vol. 4, p.1.
22. S. M. Klainer, T. B. Hirschfeld, and J. R. Golin, Paper presented at the Third Northeast Regional Meeting, American Chemical Society, October 1971.
23. T. B. Hirschfeld, Block Instrument Co. (private communication, 1973).

SUBJECT INDEX

WISWESSER LINE NOTATION INDEX*

* – Compound names in Wiswesser Line Notation (WLN) are listed in increasing alphanumeric order by columns.